Springer Tracts in Advanced Robotics

Volume 34

Editors: Bruno Siciliano · Oussama Khatib · Frans Groen

Martijn Wisse and Richard Q. van der Linde

Delft Pneumatic Bipeds

Springer

Professor Bruno Siciliano, Dipartimento di Informatica e Sistemistica, Universitá di Napoli Federico II, Via Claudio 21, 80125 Napoli, Italy, E-mail: siciliano@unina.it

Professor Oussama Khatib, Robotics Laboratory, Department of Computer Science, Stanford University, Stanford, CA 94305-9010, USA, E-mail: khatib@cs.stanford.edu

Professor Frans Groen, Department of Computer Science, Universiteit vanAmsterdam, Kruislaan 403, 1098 SJ Amsterdam, The Netherlands, E-mail: groen@science.uva.nl

Authors

Martijn Wisse
Delft University of Technology
Mechanical Engineering
Mekelweg 2
2628 CD Delft
The Netherlands
E-mail: m.wisse@tudelft.nl

Richard Q. van der Linde
Delft University of Technology
Mechanical Engineering
Mekelweg 2
2628 CD Delft
The Netherlands
E-mail: r.q.vanderlinde@tudelft.nl

ISSN print edition: 1610-7438
ISSN electronic edition: 1610-742X

ISBN 978-3-642-09191-9 e-ISBN 978-3-540-72808-5

Springer is a part of Springer Science+Business Media
springer.com

Softcover reprint of the hardcover 1st edition 2007

Foreword

At the dawn of the new millennium, robotics is undergoing a major transformation in scope and dimension. From a largely dominant industrial focus, robotics is rapidly expanding into the challenges of unstructured environments. Interacting with, assisting, serving, and exploring with humans, the emerging robots will increasingly touch people and their lives.

The goal of the new series of Springer Tracts in Advanced Robotics (STAR) is to bring, in a timely fashion, the latest advances and developments in robotics on the basis of their significance and quality. It is our hope that the wider dissemination of research developments will stimulate more exchanges and collaborations among the research community and contribute to further advancement of this rapidly growing field.

The monograph written by Martijn Wisse and Richard van der Linde is the first in the series devoted to biped robots. Research in this area has a great potential for biomechanics (understanding human locomotion), prosthetics (aiding human locomotion), and robotics (building human-like robots for entertainment and for assistance tasks). A number of pneumatic robots and bipeds based on passive dynamic walking, developed in the Delft Biorobotics Laboratory, are described in detail to understand the principles of dynamic, two-legged walking, and to use that knowledge to build human-like walking robots.

Remarkably, the monograph is based on two doctoral theses, one of which received the prize of the Fifth Edition of the EURON Georges Giralt PhD Award devoted to the best PhD thesis in Robotics in Europe. A fine addition to the Series!

Naples, Italy
December 2006

Bruno Siciliano
STAR Editor

Preface

This book reports our research on pneumatic bipeds (=two-legged robots) based on the concept of passive dynamic walking. The research was done between 1998 and 2004 at Delft University of Technology, funded by the Dutch Technology Fund STW. We, Martijn Wisse and Richard van der Linde, both obtained our PhD degrees during the project under the supervision of Prof. dr. ir. H.G. Stassen and Prof. dr. F.C.T. Van der Helm. This book is a rewritten, compacted version of the combined PhD theses.

The European robotics research community EURON awarded our research with the Georges Giralt PhD Award, which included an offer to publish the work as a monograph in the Springer Tracts on Advanced Robotics.

This book is intended for researchers who are considering to start research on two-legged robots based on passive dynamic walking. We hope that this book will convene the key ideas behind our research. We would like you to avoid our mistakes and build on our results.

This book is organized in seven chapters. The chapters 2 to 6 each present one key idea and one new robot. The background theory and computer simulations are presented where they are needed throughout the book. Some explanations may appear in more than one place so that the chapters remain mostly independent.

Many people contributed to the work that we report here. Special mention goes to Arend Schwab, our teacher of multibody dynamics and stability analysis, and to Jan van Frankenhuyzen, who designed and built the robots.

We much enjoyed our research work, and we hope that we can transmit some of that enthusiasm to you with this book.

Delft,
March 2007

Martijn Wisse
Richard Quint van der Linde

Contents

1. Introduction 1
1.1 Motivation for Biped Robot Research 1
1.2 State of the Art 3
1.3 Research Goal and Book Structure 5

2. Passive Dynamic Walking 7
2.1 Introduction 8
2.2 McGeer 8
2.3 Other Passive Dynamic Walking Research 12
2.4 Simplest Walking Model 13
2.4.1 Model 13
2.4.2 Equations of Motion 15
2.4.3 Heel Strike 16
2.4.4 Stride Function 19
2.4.5 Cyclic Motion 19
2.4.6 Results 20
2.5 Practical Hints and Tips 21

3. Baps; Pneumatic Actuation 25
3.1 McKibben Muscles 26
3.2 Pressure Control Unit 30
3.2.1 Background and Requirements 30
3.2.2 Operating Principle 30
3.2.3 Technical Realization and Results 30
3.3 Phasic Activation of a Single Pendulum Test Setup 32
3.3.1 System Model 33
3.3.2 Model Validation and Parameter Sensitivity 34
3.4 Prototype Design and Experiments 36
3.4.1 Specifications 37
3.4.2 Muscle Configuration 37
3.4.3 Feet 37
3.4.4 On-Board Control 40

3.4.5 Power Consumption 41
3.4.6 Walking Results 42

4. Mike; How to Keep from Falling Forward 45
4.1 Introduction 46
4.2 Modeling and Analysis 46
4.2.1 The Simplest Walking Model 46
4.2.2 Basin of Attraction 47
4.2.3 Failure Modes 47
4.2.4 Behavior Inside the Basin of Attraction 48
4.2.5 Basin of Attraction Versus Slope Angle 50
4.2.6 Basin of Attraction Versus Eigenvalues 52
4.3 Swing Leg Action for a Larger Basin of Attraction 52
4.3.1 Largest Possible Basin of Attraction 52
4.3.2 The Rimless Wheel 54
4.3.3 A Realistic Actuation Model 55
4.4 Prototype Design and Experiments 57
4.4.1 Prototype Construction 57
4.4.2 Actuation System 57
4.4.3 Stability Results 58
4.5 Discussion 59
4.5.1 Level Floor 59
4.5.2 Distributed Leg Mass 60
4.5.3 Feet 60
4.5.4 Knees and Muscles 60
4.5.5 Human Walking 61
4.6 Conclusion 61

5. Max; Adding the Upper Body 63
5.1 Introduction 64
5.2 Point Mass Model with Upper Body 64
5.3 Results of Point Mass Model 67
5.3.1 Walking Motion 67
5.3.2 Inherent Stability 68
5.4 Parameter Study on Point Mass Model 70
5.4.1 Slope and Spring Stiffness; Speed and Step Length 70
5.4.2 Upper Body Height and Weight 71
5.4.3 Limits to Stability 71
5.5 Prototype Design 72
5.5.1 Simulation Model 72
5.5.2 Simulation Procedure 73
5.5.3 Default Parameter Values 74
5.5.4 Construction of the Prototype 74
5.6 Prototype Experiments 77
5.6.1 Resultant Motion and Gait Characteristics 77
5.6.2 Stability 79

5.6.3 Parameter Sensitivity ... 81
5.6.4 Energy Efficiency ... 82
5.7 Conclusion ... 83

6. Denise; Sideways Stability ... 85
6.1 Introduction ... 86
6.2 Advantageous Lean-to-Yaw Coupling ... 86
6.2.1 Concept ... 86
6.2.2 Skateboard ... 87
6.2.3 Bicycle ... 88
6.3 Simplest Passive Walking Model with Lean-to-Yaw Coupling ... 91
6.3.1 Model ... 91
6.3.2 Equations of Motion ... 93
6.3.3 Simulation Procedure ... 94
6.4 Simulation Results ... 95
6.4.1 Fully Passive Model ... 95
6.4.2 Model with Hip Actuation ... 98
6.4.3 Stability Versus Velocity ... 100
6.4.4 Walking and Steering ... 103
6.5 Applicability in Real Walking Robots ... 105
6.6 Prototype Design ... 105
6.7 Prototype Experiments ... 109
6.7.1 Gait Characteristics ... 109
6.7.2 Motions ... 111
6.7.3 Energy Consumption ... 113
6.8 Conclusion ... 114

7. Discussion and Future Research Directions ... 117
7.1 On General Design Guidelines for Stability ... 117
7.2 On Stability Measures and Disturbances ... 120
7.3 On Foot Contact in Simulations ... 122
7.4 On Human Walking ... 123
7.5 Future Directions ... 125

References ... 127

1. Introduction

1.1 Motivation for Biped Robot Research

This is a book about biped robots. Why would someone research biped robots? In ancient Dutch, the answer would be: 'Ter leering ende vermaeck', which means 'for learning and for fun'. From the research, we can learn the principles of dynamic two-legged walking. This knowledge can be used for biomechanics (understanding human locomotion), prosthetics (aiding human locomotion), and robotics (building human-like robots for entertainment and for assistance tasks).

Biomechanics

By building robots, we can learn more about human locomotion. The approach of robot construction complements the usual approach of studying actual human beings, i.e. the area of *gait analysis*. Let's first review what has been done in that area, and then provide an argument in favor of building robots.

The human gait has been studied since early history[1], such as Hippocrates' [73] treatment of people with a hip joint dislocation around 400 BC. The early modern times have produced beautiful treatments on human motion (e.g., Borelli [16] in 1680). As soon as photography became available in the late nineteenth century, Marey [103] and Muybridge [118] applied it to perform two-dimensional (2D) kinematic analyses. The first 3D kinematic analysis was performed in 1891 by Braüne and Fischer [18]. Elftman [47] developed the force plate in 1938 which he used to perform the first inverse-dynamics gait analysis. Although the tools have improved enormously since then (e.g. with the introduction of the 3D computerized data acquisition and even real-time inverse-dynamics analysis), the basic analytical approach has remained the same ever since; the motions and ground reaction forces are measured and the joint torques or muscle forces are then calculated using a model of the human body.

[1] The remarks with respect to the history of gait analysis and prosthetics in this chapter are based on the *Clinical Gait Analysis* webpages, `http://www.univie.ac.at/cga/`

M. Wisse and R.Q. van der Linde: Delft Pneumatic Bipeds, STAR 34, pp. 1–5, 2007.
springerlink.com

Although human gait analysis is indispensable, we believe that by itself it is not sufficient for creating an understanding of the principles of dynamic walking. First, it is hard to discriminate between essential and non-essential features, because the system has to be studied in its entirety. Second, some essential features might be overlooked because of their sustained presence in all experiments. For example, many studies based on inverse-dynamics calculations ignore the problem of stability and thus fail to recognize the importance of related features such as local feedback control loops.

We believe that the best way to gain insight is to build up an artificial walking system. By adding one feature (e.g. hip actuation, or an upper body) at a time, we ensure a focus on the essential features for successful walking. For each additional feature it is known exactly why it is necessary and how it contributes to human-like walking. This is the approach taken throughout the research presented in this book.

Prosthetics

There are two types of artificial limbs: prostheses (limb replacement) and orthoses (limb function support). Artificial limbs have two purposes: to restore the function of the limb (prostheses and orthoses) and to hide the weakness or disfiguredness of a missing limb (mostly prostheses). In ancient history, most of the amputations were a result of human conflicts, and this is not different in the more recent history, as World Wars I and II brought tens of thousands of amputees each, while other parts of the world continue to suffer from large scale conflicts. In addition, the late 1950's saw the 'Thalidomide tragedy'; approximately ten thousand babies were born with deformed or missing limbs due to the disastrous side effects of this insomniac drug taken during pregnancy[2]. In the Netherlands, currently most amputations are a result of an accident or are necessary because of a vascular disease (mostly elderly patients), and add up to over 2000 amputations per year in the Netherlands alone [95], a country with 17 million inhabitants.

For most part of history, leg protheses have been constructed the same way, with a wooden peg leg, leather straps and a soft leather or linen lining. In the sixteenth century Paré heralded an era of mechanical refinement comprising better materials and more degrees of freedom, especially for the upper extremities [128]. For the legs, however, even the most advanced prostheses still have only one degree of freedom, the knee. Foot designs have evolved to incorporate damping and compliance, but most of these developments were experience-based. The same is true for the modern prostheses with computer controlled damping in the knee. Almost none of the design features of today's prostheses are based on knowledge of the dynamics of walking. This lack of knowledge results in unnatural dynamic behavior of the prosthetic leg, which the amputee will try to hide by means of

[2] The 'Thalidomide tragedy' ('Softenon drama' in Dutch) was one of the incentives to start research into prosthetics at Delft University of Technology, the technology of which has been the basis for our research on biped robots.

extra effort [142]. Similarly, gait orthoses are usually fitted and tuned based on experience rather than on sound knowledge about dynamic walking. We hope that a better understanding of the principles of dynamic walking will aid the development of more comfortable (more 'natural') prostheses and orthoses.

Robotics and Entertainment

The field of entertainment has less urgency but much more economic thrust than the field of rehabilitation. The billion dollar markets of computer games and motion pictures make more and more use of computer generated actors [76]. The generated motions must be of high quality because the human eye is very perceptive for deviations from natural walking motions. Although virtual gravity is a little more forgiving than the real thing, knowledge of the underlying dynamics is imperative for the development of realistic animations.

Recently, the entertainment industry has been opening up the market for entertainment robots. After SONY's four-legged AIBO-dog, several companies [136, 100, 86, 99] are now developing two-legged *humanoid* robots. The attractiveness of the human appearance of these first generation humanoid robots has already resulted in huge media coverage and public interest, providing the developing companies with an effective means to show their technological disposition. It is expected that the human-like means of locomotion will eventually also prove useful for tasks other than entertainment; a two-legged design provides functionality in environments that are especially designed for humans such as dangerous factory environments, construction workplaces, and the homes of private robot owners. Various reports [148, 134] predict a steady development from the current research and entertainment products towards highly versatile machines, parallel to the stages of development seen in the short history of the PC. For the development of two-legged walking robots, the need for knowledge of the principles of dynamic human-like walking is self-evident.

1.2 State of the Art

The research in this book does not stand alone. On the contrary, two-legged robots are becoming ever more popular as a research topic. Here we will present an overview of the various projects at the time of writing (2006). The order of the presentation is loosely based on the 'degree of passivity' of the walking robots; on one end we find the *static bipeds* which have a slow, 'robotic-looking' gait, and on the other end we find the *passive dynamic bipeds* which have an efficient gait in a natural cadence. Our robots are based on the passive dynamic bipeds which we will treat in detail in the next chapter. The story in this section starts from the other end; the static bipeds.

The standard approach to robot control as applied in industrial manufacturing robots is to ensure direct control over all degrees of freedom. By application of strong actuators and stiff structural components, the robot manipulators can be controlled to accurately track prescribed trajectories. This approach has been used as a starting point for the construction and control of biped robots. To deal with the fundamental characteristics of limited foot contact and discretely changing dynamics, the bipeds are programmed to walk slowly and to keep the center of mass above the foot contact area (this is called 'static walking'). Thanks to these constraints, the stance foot remains in full, flat contact with the floor and thus it 'simulates' direct control over all degrees of freedom.

An extension towards a more dynamic approach is the so-called 'Zero Moment Point' [169] (center of pressure [60]) control; by keeping the center of pressure inside the foot contact, full and flat foot contact can also be ensured. Note that this is not sufficient to prevent the biped from falling, it would just fall while keeping full foot contact. Usually, the 'Zero Moment Point' calculation serves as a constraint to trajectory generators so that faster (than static) walking motions can be generated while ensuring that the foot maintains full, flat contact with the floor. The actual stability of the gait results from the ability to robustly track the generated trajectory. One of the first research robots based on this approach was built at Waseda University in 1970 [5]. Many followed, especially in Japan, and work towards a more dynamic gait is in progress.

The benefit of the standard industrial approach is that a complete system can be built from the start; it has all the degrees of freedom needed to make it look human and thus it is immediately ready for commercialization, e.g. for entertainment purposes. Making it more dynamic, natural, efficient and cheaper are topics for gradual improvement, as illustrated with the succession of prototypes by Honda [74, 136] and Sony [100]. This benefit makes this approach a good starting point for industrial developers. The drawback, as we see it, is that these robots are over-controlled and thus may consume more energy than necessary. For the task of walking without falling, it is certainly not necessary to exactly track each pre-determined joint trajectory all the time.

Successful walking has also been achieved with more biologically inspired approaches. Various robots have shown successful gait being controlled by Central Pattern Generators [48, 57, 135], or being controlled more intuitively using Virtual Model Control [130] and various other techniques [75, 64, 85].

Even more related to passive dynamic walking are those projects that used passive ankles or arc feet or point feet. In this category we find the early work by Katoh [88], the Hybrid Zero Dynamics approach [65, 66, 171] and the related Passive Dynamic Autonomous Control [131]. Various learning algorithms [116, 114, 117, 120, 149] have led to quick learning on arc feet robots. Finally, the Self Excited Biped Locomotion [123, 124] is with only one hip motor perhaps the closest relative of Passive Dynamic Walking.

The robots that we present in this book are powered by pneumatics. Although this is somewhat uncommon (all bipeds mentioned above have electric actuators), we are not the only group that applies pneumatics. Interestingly, most of the pneumatically powered bipeds (including our own) use pneumatic 'muscles' rather than piston-type actuators. One of the first to incorporate pneumatics is the Japanese pioneer for legged locomotion Kato. During the 1960s and the 1970s he has built several statically balanced walking bipeds such as WAP I,

WAP II and WAP III [87]. In Paris the pneumatic biped Bipman [68] has been developed at INSERM, and it now resides at the University of Toulon. At the Laboratoire de Robotique de Versailles, the pneumatic biped STEP [119] was built with pneumatic cylinders. Caldwell, at the University of Salford, developed the biped Salford Lady [9]. From 1987 to 2005 the Shadow Company [6] won several awards with their inspiring McKibben muscle powered biped. Since the mid 1990's, Vrije Universiteit Brussel has been working on an alternative for McKibben muscles, Pleated Pneumatic Artificial Muscles [41, 166]. These muscles are used to power the biped Lucy [165, 167, 163, 164] which is controlled according to the trajectory tracking approach. Finally, as a follow-up to the research presented in this book, Osaka University is now building a series of bipeds powered with McKibben muscles [146, 145, 147, 144, 77].

1.3 Research Goal and Book Structure

Our final research goal is to understand the principles of dynamic, two-legged walking, and to use that knowledge to build human-like walking robots. During the period described in this book (1998-2004), we have made significant steps toward that goal. We have mastered 'Passive Dynamic Walking' (Chapter 2), developed a pneumatic actuation system (Chapter 3), investigated how to keep from falling forward (Chapter 4), added an upper body (Chapter 5), and investigated 3D stability (Chapter 6). In Chapter 7, we will discuss future research directions. All prototypes presented in this book have demonstrated successful walking motions. Movies can be found at our website [1].

2. Passive Dynamic Walking

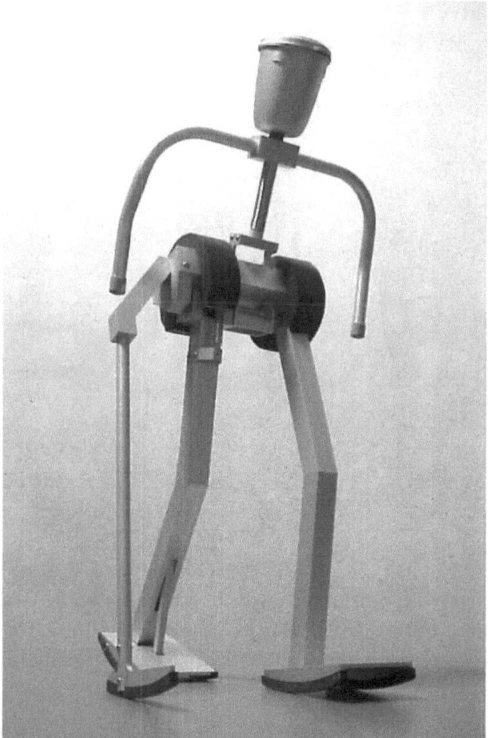

M. Wisse and R.Q. van der Linde: Delft Pneumatic Bipeds, STAR 34, pp. 7–24, 2007.
springerlink.com

2.1 Introduction

The concept of passive dynamic walking is the starting point for our research. In this chapter, we would like to review the work that has been done on passive dynamic walking. Although we have made some contributions of our own to this field, most of the research reported in this chapter was done by others.

Long before McGeer introduced the term 'Passive Dynamic Walking' in 1987, there have been several hints towards the possible role of passive dynamic motions in human walking. A remarkably relevant hypothesis posed by Weber and Weber [170] as early as 1836 reads: 'Die Beine können am Rumpfe wie Pendel hin und her schwingen. (...) Unsere aufmerksamkeit wird für diese schwingende Bewegung nicht erfordert.' ('The leg can swing back and forth like a pendulum suspended from the body. ... Our attention is not required to produce this swinging motion.') Mochon and McMahon [113] arrived at the same conclusion after comparing the swing leg motion with a passive double pendulum. Another hint in that direction is given by Ralston [133] who discovered that there exists an optimal walking velocity for humans; at approximately 5 km/h the cost of transport (i.e. energy cost per weight per distance traveled) is minimal, a phenomenon that indicates the use of the natural frequencies of the mechanical system.

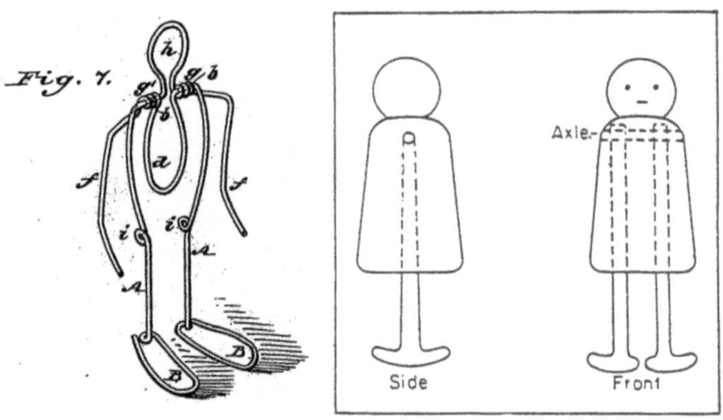

Fig. 2.1. Toy designs from old patents [thanks to Ruina for collecting such patents]. **Left:** Wireframe toy design by Fallis, 1888. **Right:** Passive walking toy by Wilson ('Wilson Walkie'), 1938. Both toys require a mild downward slope for a sustained walking motion.

Early toy makers [172, 50] (Fig. 2.1) proved the applicability of the ideas by showing that the human walking motion can at least partially be generated with passive mechanisms that move and oscillate at their natural frequencies.

2.2 McGeer

In 1987, McGeer started working on the concept of passive dynamic walking. He wrote a number of ground-breaking papers about this research [106, 108,

109, 107, 110, 111, 112]. Here we try to provide an overview of his work, but the reader is strongly recommended to review McGeer's original papers. At the time of writing, a complete collection of these papers is being maintained at the website of prof. Ruina at Cornell University [2].

McGeer's first contribution was the conception of two elementary 2D models; the 'Rimless Wheel' (following [104]) and the 'Synthetic Wheel'. Each model captures one basic aspect of passive dynamic walking.

- **The rimless wheel** captures heel strike impacts and demonstrates the basic stabilizing mechanism. It is an imaginary cart-wheel without rim (Fig. 2.2), which bumps down a shallow slope. At each step, some energy is gained from gravity, but this is lost again in the impact when the next spoke makes contact. If there is a balance between energy input and energy loss, the system is in a steady limit cycle. If there is an unbalance, this will be automatically restored. For example, if we push the rimless wheel to a higher velocity, it will make more violent impacts and thus lose more energy than usual, while the energy gain per step remains the same. The speed v will asymptotically go back to the equilibrium speed v_0. The convergence rate is easy to calculate, especially if we assume that all the mass is concentrated as a point mass at the hip (i.e. the wheel hub). In that case, for an inter-spoke angle of ϕ (see Fig. 2.2), the convergence is:

$$\Delta v_{n+1} = \Delta v_n \cos^2 \phi \tag{2.1}$$

  An extensive treatment of the rimless wheel can be found in [30] (2D analysis) and [31] (3D analysis).

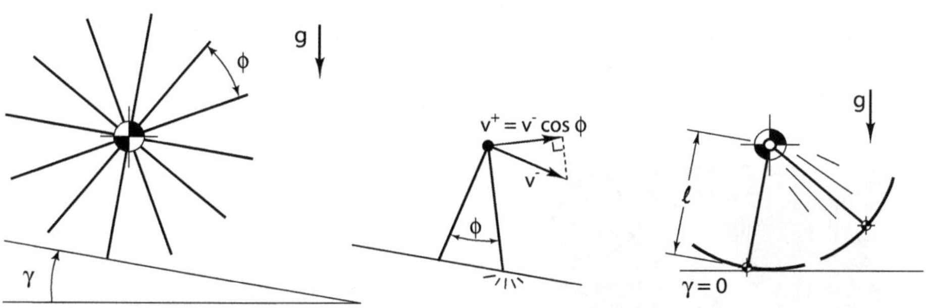

Fig. 2.2. Left: Rimless wheel. **Middle:** Impact equation for rimless wheel. **Right:** Synthetic wheel.

- **The synthetic wheel** (Fig. 2.2) captures the passive motion of the swing leg. In this model, the rim is not removed, but cut halfway between each spoke. All but two spokes are removed, and a pin joint is put at the hub. Now we have two legs with big semicircular feet, which roll forward on a level floor like a 'continuous' wheel composed of alternating legs. We assume again that there is a big point mass at the hip joint, whereas the legs are infinitesimally lightweight. This makes the forward rolling motion of the stance leg independent of the passive swinging motion of the other leg. The infinitesimally small leg mass is concentrated as a point mass at the rim, determining the natural frequency at $\sqrt{g/l}$ with gravity g and leg length l, see Fig. 2.2. The swing leg swings forward like a passive pendulum at the same time that the stance leg rolls forward with constant speed. For each speed, one can find appropriate initial conditions so that the step ends exactly how it begun (but with the swing and stance legs reversed). The most important result is that the step time turns out to be approximately 2/3 of the pendulum period, i.e. $\frac{2}{3} \cdot 2\pi\sqrt{l/g}$, independent of the forward velocity. In other words, there is an (energetically) optimal gait frequency which is determined by the hardware parameters. By keeping the hip joint unactuated, the walker cannot show any other gait than the most efficient one. As efficiency was perhaps his main target, McGeer used a passive hip joint in all of his models and prototypes.

Next, McGeer simulated and built a straight-legged passive walker, see Fig. 2.3. It has 2D dynamics thanks to the double leg pairs (the outer legs form a unit and the inner legs form a second unit). In this book, we always use this solution for 2D prototypes, as opposed to the less autonomous solution of walking in circles by means of attachment to a boom (as used in e.g. [130, 132, 171]). The arc feet of McGeer's prototype (Fig. 2.3) are another design idea that we use throughout this book. He uses a radius of approximately 1/3 of the leg length, as it is a human-like value [70] and also it provides the best stability results. The prototype is thus somewhere between the rimless wheel and the synthetic wheel. The straight legs bring a small problem of 'foot-scuffing' at midstance; the prototype has no knees or lateral motion to make clearance for the swing foot. Initial trials were therefore done with tiles (stepping stones) on the floor, creating effective swing space but forcing a preemptive decision on the positions

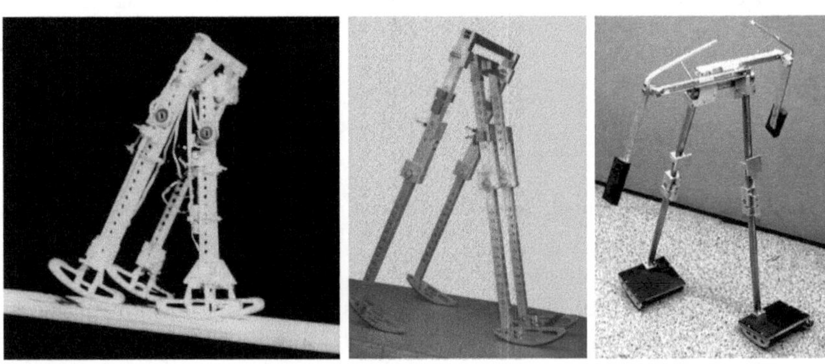

Fig. 2.3. Left: McGeer's straight legged passive dynamic prototype. **Middle:** Cornell University's copy [52] of McGeer's passive dynamic walker with knees. **Right:** Cornell University's 3D passive dynamic walker by Collins, Wisse, and Ruina [36].

of the footprints. As an alternative solution, the prototype in Fig. 2.3 has a lead screw mechanism on each leg that folds the swing foot sideways for clearance at midstance. With this prototype and the associated model, McGeer validated the concept of passive dynamic walking. He then researched the effects of foot radius and mass distribution, noting that especially the fore-aft mass balance is a delicate parameter. Another interesting result concerns the vertical mass distribution. The center of mass (CoM) should be rather close to the hip, as opposed to a statically stable system which would required the CoM to be located as low as possible. Finally, he found that friction in the hip joint is unfavorable for passive walking, but its negative effect can be compensated by moving the CoM of the legs a few millimeters backward.

McGeer continued by creating a simulation model and prototype with knees, see Fig. 2.3. The knees bend and then extend passively during the swing phase until a hyper-extension stop is reached. An intentionally leaky suction cup prevents bouncing during this knee strike. Due to the leak, the suction cup does not resist the knee motion for the next swing phase. After knee strike, the knee remains fully extended during the remainder of the swing phase. Then, during the heel strike impact, and during the subsequent stance phase, the knee remains extended because the ground reaction force pushes the knee in its end stop. This requires the feet to be mounted in a forward position, as opposed to the straight-legged prototype which has symmetrically mounted feet (Fig. 2.3). With this prototype, McGeer clearly demonstrated the applicability of passive dynamic walking as a model for human walking. The gait looks superbly natural, requires no controls and uses hardly any energy (less than 0.05 Joule per unit of robot weight per meter traveled). A parameter study revealed that the mass of the shank should be high up, near the knee, and the mass of the thigh should be close to the hip joint.

Although famous for passive walking, McGeer actually has always been a strong advocate of powered walking. Throughout his work, one can find suggestions as to how the downhill slope can be replaced by periodically 'pumping the gait', for example through push-off. Powered walking really formed the core motivation for his research, judging by his reference to the Wright Brother's success in powered flight. They first mastered the aerodynamics and control of passive gliding, after which the addition of the engine was a minor modification. Thus, passive walkers are the 'gliders' amongst the walking robots. McGeer envisioned that by using passive dynamics as the basis of the walking robot design, it must be possible to create highly efficient powered walking robots. In this book, we adhere to this vision, by adding minimal amounts of power to otherwise passive robots.

After the kneed prototype, McGeer performed a range of simulation studies on running, 3D walking, models with an upper body, uphill walking, etcetera. Many ideas are still highly relevant and are waiting for someone to execute them. McGeer, in the mean time, has moved away from walking research back to his aviation roots, and is now building unmanned aircraft [3]. Nevertheless, his interest in the bipeds remains, witness his occasional appearance at conferences on dynamic walking and especially the recent announcement of his 'W-prize' for the first efficient robot that can complete a 10 kilometer long obstacle course [7].

2.3 Other Passive Dynamic Walking Research

There are only a few researchers after McGeer that have built fully passive prototypes. Most of it happened at the Ruina's group at Cornell University, who started by building, simulating, and testing a copy of McGeer's kneed prototype [52]. In the same group, Coleman discovered a 3D passive walker design that cannot stand stably, yet it is stable when it walks [33, 32]. Also at Ruina's group, Collins and Wisse built a 3D walker with knees and counter-swinging arms [36]. The design was based on McGeer's 2D kneed prototype combined with the foot design of an old toy patent [50]. This prototype was the first 3D passive walker with reasonable human-like motions, and as such the robot has been very helpful in selling the concept of passive dynamic walking. Recently, the Fujimoto lab at Nagoya [82] succeeded in obtaining a 15 minute continuous passive walk on a downward tilted treadmill. Other researchers that have recently published on passive prototypes include Mayer [105], and Tedrake [150] who built a 3D straight legged walker much like the original 'Wilson walkie' [172] later to be used for machine learning of walking [149].

In our group, we have built two fully passive walkers (Fig. 2.4). The first one is a reincarnation of McGeer's straight-legged walkers using a floor with tiles for foot clearance. We used it to get acquainted with passive dynamic walking,

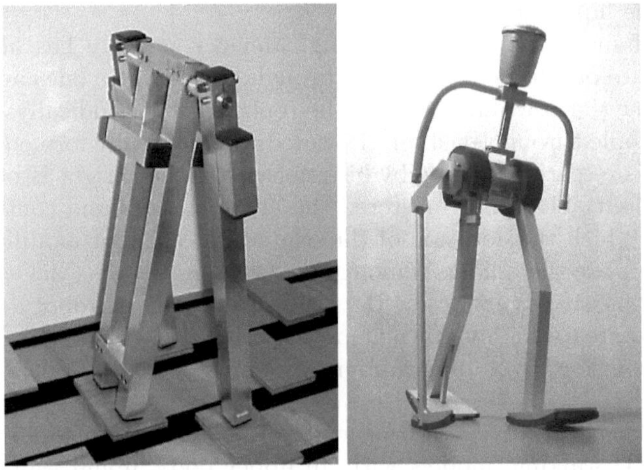

Fig. 2.4. Left: Our version of McGeer's straight legged passive dynamic prototype, walking on a floor with tiles. **Right:** Our 'Museon' walker, more robust and appealing but topologically identical to the left prototype, except for the addition of a mechanism that lifts the middle foot during the swing phase, so that no floor tiles are required.

and to find that a larger foot radius leads to better disturbance handling (later theoretically analyzed in [179]). Our second passive walker (Fig. 2.4) is topologically the same. It has only one internal degree of freedom which is the hip joint. We played with the shapes of the legs, and attached a lightweight body to the inner leg, in order to give it a more appealing appearance. This robot was built for demonstration purposes, with a sturdy construction ensuring the survival of many falls after launches by untrained members of the audience. This robot has one extra feature when compared to the first prototype, namely a small mechanism that lifts the ankle of the middle leg. Thanks to this mechanism, which was inspired by an earlier prototype from Cornell University, the prototype does not need the tiles on the walking surface.

Much more literature is available on simulation studies of passive dynamic walking models. A key development from Ruina's group was Garcia *et al.*'s introduction of the 'Simplest Walking Model' [54], which we will present in detail in Section 2.3. At the same time, Goswami presented a similar model [61, 62]. More extended models (with knees and 3D dynamics) were studied by Dankowicz [8], and by Wisse and Schwab [181]. Early on, Kuo [96, 97, 98] recognized passive dynamic walking as a highly useful abstraction of human walking for biomechanical studies. In the world of control science, Spong [143] and Asano [10] independently used the simplest walking model as a basis for a control law that could simulate passive downhill walking while actually walking on level ground or even uphill. Also some efforts have been reported on learning controllers that use a passive model and a Poincaré Map stability analysis as the basis [115, 116]. In addition to these studies on passive walking models, there are many simulation studies with neural networks or Central Pattern Generator controllers and the like, but these are too numerous to present here, as we try to focus on those studies that explicitly use a Poincaré Map analysis for stability assessment. Since the well-attended 'Dynamic Walking' workshop in May 2006 [4], we expect to see many more research projects on passive(-based) dynamic walking.

2.4 Simplest Walking Model

In all of the passive dynamic walking studies, the gait stability is assessed with the Poincaré Mapping analysis. In 1998, Garcia *et al.* wrote an excellent paper [54] that was meant as a tutorial for such simulation studies. The paper features a 2D straight-legged walking model. In this book, we use this model often as a point of departure for our analyses. Therefore, we would like to repeat Garcia *et al.*'s tutorial here, with a modified derivation of the equations of motion and impact [141]. Note that the text is based on one of our papers [141] but all of the results at the end of this section are repeated from Garcia *et al.* [54].

2.4.1 Model

The model, shown in Fig. 2.5, consists of two rigid links with unit length, connected by a frictionless hinge at the hip. The mass is distributed over three point masses; one with unit mass at the hip, and two with mass m_f at the feet. The limit case where the foot mass is negligible in comparison with the hip mass, $m_f \to 0$, is investigated. This unactuated two-link system walks down a slope in a gravity force field with unit magnitude. The scaled model of the walker now only has one free parameter, the slope angle γ.

A walking step is started with both feet on the slope. The front foot has just made ground contact, the hind foot has a velocity away from the floor. During a step, the stance foot is modeled as a hinge, connected to the floor. The swing foot is moving freely as the other end of a double pendulum. At about midstance, the swing foot will briefly be below floor level ('foot-scuffing'), which is inevitable for a walker with straight legs. Knees ([109], [175], [8], [36]) or other leg shortening measures ([157]), as well as 3D motion ([96], [157] [159]) would solve the problem but increase complexity of the model. After this short through-pass, the second time that the swing foot reaches floor level is regarded as heel-strike, the end of the step. The former swing foot makes a fully inelastic collision and becomes the new stance leg. Instantaneously, the former stance leg loses ground contact, and a new step begins.

For the analysis of the passive dynamic walking motion, we will have to apply the following methods. First, we must derive the equations of motion for the

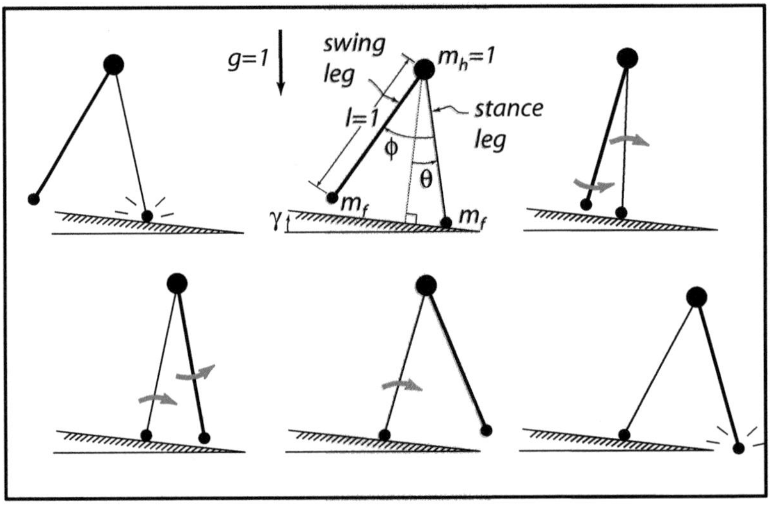

Fig. 2.5. A typical passive walking step. The new stance leg (lighter line) has just made contact with the ramp in the upper left picture. The swing leg (heavier line) swings until the next heelstrike (bottom right picture). The top-center picture gives a description of the variables and parameters that we use. θ is the angle of the stance leg with respect to the slope normal. ϕ is the angle between the stance leg and the swing leg, m_h is the hip mass, m_f is the infinitesimally small foot mass, l is the leg length, γ is the ramp slope, and g is the acceleration due to gravity. Reprinted with permission from Garcia *et al.* [54].

walker during the support phase. They will then be solved by numeric integration. Second, we formulate and apply the impact equations governing the heel strike. Third and last, we will formulate the support exchange and combine the results from the previous stages in what McGeer calls a 'stride function'. The stride function [110] is a Poincaré map relating the state during one part of a step with the state during the same part of the next step. Finally, we will find cyclic walking motions as solutions to the stride function.

2.4.2 Equations of Motion

The method that we present here for the derivation of the equations of motion is based on the concept of 'virtual work' as we will explain. The resulting equations are equal to those derived with Lagrange's method, but our method is simpler in execution and thus less error-prone.

The configuration of the walker is defined by the coordinates of the three point masses; the stance foot, the hip and the swing foot, which can be arranged in a global vector $\mathbf{x} = (x_{stl}, y_{stl}, x_{hip}, y_{hip}, x_{swl}, y_{swl})^T$. These coordinates are not independent owing to the two distance constraints imposed by the stance and the swing leg. In order to eliminate the constraint forces from the start, we express the equations of motion in terms of independent generalized coordinates. Let Oxy be a fixed orthogonal system of coordinates with Ox along the walking slope and Oy directed upward normal to the slope. Then u and v are the coordinates of the contact point of the stance foot. During walking motion they will be fixed, but at heel-strike they will be free so that they don't participate in the impact equation. Furthermore, θ is the absolute angle of the the stance leg with respect to Oy, and ϕ is the relative angle between the swing leg and the stance leg. The configuration of the walker can be described by the vector of generalized coordinates $\mathbf{q} = (u, v, \theta, \phi)^T$. The coordinates $\mathbf{x}$ can locally be expressed as functions of the generalized coordinates $\mathbf{q}$, the kinematic degrees of freedom (configuration coordinates), by means of a transfer function $\mathbf{F}$ as

$$\mathbf{x} = \mathbf{F}(\mathbf{q}) \rightarrow \begin{bmatrix} x_{stl} \\ y_{stl} \\ x_{hip} \\ y_{hip} \\ x_{swl} \\ y_{swl} \end{bmatrix} = \begin{bmatrix} u \\ v \\ u - \sin(\theta) \\ v + \cos(\theta) \\ u - \sin(\theta) + \sin(\theta - \phi) \\ v + \cos(\theta) - \cos(\theta - \phi) \end{bmatrix}. \tag{2.2}$$

The unreduced equations of motion for the system are obtained by assembling the contribution to the virtual power equation of all point masses in a global mass matrix $\mathbf{M}$ and a global force vector $\mathbf{f}$, which results in a virtual power balance

$$\delta\dot{\mathbf{x}}^T[\mathbf{f} - \mathbf{M}\ddot{\mathbf{x}}] = 0. \tag{2.3}$$

Here, $\delta\dot{\mathbf{x}}$ are kinematically admissible virtual velocities, which satisfy all instantaneous kinematic constraints. By differentiating the transfer function (2.2) we obtain

$$\dot{\mathbf{x}} = \mathbf{F}_{,\mathbf{q}}\dot{\mathbf{q}}, \quad \delta\dot{\mathbf{x}} = \mathbf{F}_{,\mathbf{q}}\delta\dot{\mathbf{q}} \quad \text{and} \quad \ddot{\mathbf{x}} = \mathbf{F}_{,\mathbf{q}}\ddot{\mathbf{q}} + \mathbf{F}_{,\mathbf{qq}}\dot{\mathbf{q}}\dot{\mathbf{q}}. \tag{2.4}$$

Here a subscript comma followed by one or more variables denotes partial derivatives with respect to these variables. The way in which higher-order derivatives have to be multiplied by the juxtaposed vectors goes without saying. Substitution of these expressions in the virtual power equation (2.3) and adding on the left-hand side the contribution, $\delta\dot{\mathbf{q}}^T\mathbf{Q}$, from the generalized forces $\mathbf{Q}$ dual to the coordinates $\mathbf{q}$, yields the reduced equations of motion

$$[\mathbf{F}_{,\mathbf{q}}^T\mathbf{M}\mathbf{F}_{,\mathbf{q}}]\ddot{\mathbf{q}} = \mathbf{F}_{,\mathbf{q}}^T[\mathbf{f} - \mathbf{M}\mathbf{F}_{,\mathbf{qq}}\dot{\mathbf{q}}\dot{\mathbf{q}}] + \mathbf{Q}. \tag{2.5}$$

For the walker the global mass matrix is

$$\mathbf{M} = \text{Diag}(m_f, m_f, 1, 1, m_f, m_f), \tag{2.6}$$

and the applied forces, only gravity, are

$$\mathbf{f} = \mathbf{M}[\sin(\gamma), -\cos(\gamma), \sin(\gamma), -\cos(\gamma), \sin(\gamma), -\cos(\gamma)]^{\mathrm{T}}, \tag{2.7}$$

and zero for the generalized forces Q_θ and Q_ϕ. The fact that the stance foot is in contact with the floor results in the boundary conditions $u = 0$ and $v = 0$. This contact is only valid for compressive vertical contact force, $Q_v > 0$, and will be checked during the simulation. After solving the unknown accelerations of the generalized coordinates $\ddot{\mathbf{q}}$ from the reduced equations of motion (2.5) and then taking the limit of infinitesimally small foot mass yields

$$\lim_{m_f \to 0} \ddot{\mathbf{q}} = \begin{bmatrix} \ddot{\theta} \\ \ddot{\phi} \end{bmatrix} = \begin{bmatrix} \sin(\theta - \gamma) \\ \sin(\phi)(\dot{\theta}^2 - \cos(\theta - \gamma)) + \sin(\theta - \gamma) \end{bmatrix}, \tag{2.8}$$

Note that we have omitted the solutions for u and v, because they are identical to zero. For the unknown contact forces the solution is

$$\lim_{m_f \to 0} \begin{bmatrix} Q_u \\ Q_v \end{bmatrix} = \begin{bmatrix} \sin(\theta)(\dot{\theta}^2 - \cos(\theta - \gamma)) \\ -\cos(\theta)(\dot{\theta}^2 - \cos(\theta - \gamma)) \end{bmatrix}. \tag{2.9}$$

In the case of a more complicated walker, as for example in the 3D passive dynamic biped with yaw and roll compensation [175], it will be impractical to solve symbolically for the accelerations of generalized coordinates. In that case, it is more practical to do a numerical evaluation of every individual contribution to the reduced equations of motion (2.5) and its solution.

2.4.3 Heel Strike

We assume that the heel strike behaves as a fully inelastic impact (no slip, no bounce), which is in accordance with observations on existing passive dynamic walking prototypes. Also, double stance is assumed to occur instantaneously. As soon as the swing foot hits the floor the stance foot lifts up, not interacting with the ground during impact. The resulting vertical velocity of the lifting foot should then be pointed upward. If this is confirmed after the impact equations are solved, the assumption is self-consistent [23]. Otherwise, the walker would come to a complete stop.

Treating heel strike as an impact, we assume that velocities change instantaneously. These velocity jumps are enforced by very high values of the contact forces acting only during a small time interval of contact. In the limit case the forces go to infinity and the time interval goes to zero. The integral of the force with respect to time over the duration of the impact, the impulse, has a finite value which is the cause of the velocity jump. While the impact takes place all positions as well as all non-impulsive forces of the system remain constant. The impact is usually divided into a compression and an expansion phase. Newton's impact law links these two phases by stating that the relative speed after impact equals e times the relative speed before impact but in opposite direction. The factor e is the coefficient of restitution. A value of $e = 1$ corresponds with a fully elastic impact whereas the value of $e = 0$ represents a completely inelastic impact in which the two parts "stick" together after impact. The reduced equations of motion (2.5) written in terms of the generalized coordinates $\mathbf{q}$ are

$$\bar{\mathbf{M}}\ddot{\mathbf{q}} = \bar{\mathbf{f}}, \tag{2.10}$$

with the reduced mass matrix and force vector

$$\bar{\mathbf{M}} = [\mathbf{F}_{,\mathbf{q}}^T \mathbf{M} \mathbf{F}_{,\mathbf{q}}], \quad \bar{\mathbf{f}} = \mathbf{F}_{,\mathbf{q}}^T [\mathbf{f} - \mathbf{M} \mathbf{F}_{,\mathbf{qq}} \dot{\mathbf{q}} \dot{\mathbf{q}}] + \mathbf{Q}. \tag{2.11}$$

Note that the 'lifting stance foot'-assumption implies that the system has no boundary conditions on the former stance foot and consequently there are more degrees of freedom during impact than during smooth motion. The uni-lateral constraints at heel strike are expressed by the contact functions $\mathbf{g}$, the coordinates of the swing foot expressed in terms of the generalized coordinates as

$$\mathbf{g}(\mathbf{q}) = \begin{bmatrix} g_x \\ g_y \end{bmatrix} = \begin{bmatrix} x_{swl} \\ y_{swl} \end{bmatrix} = \begin{bmatrix} u - \sin(\theta) + \sin(\theta - \phi) \\ v + \cos(\theta) - \cos(\theta - \phi) \end{bmatrix}. \tag{2.12}$$

When contact occurs, detected by a change of sign in the swing foot vertical clearance function g_y, the former swing foot becomes constrained in both the x and y direction and the equations of motion become

$$\bar{\mathbf{M}}\ddot{\mathbf{q}} + \mathbf{g}_{,\mathbf{q}}^T \boldsymbol{\lambda} = \bar{\mathbf{f}}, \tag{2.13}$$

with the Lagrangian multipliers $\boldsymbol{\lambda}$ dual to the relative contact velocities $\dot{\mathbf{g}}$. These multipliers can be interpreted as the contact forces. Integration of these equations of motion over the time of impact and taking the limit case yields

$$\lim_{t^- \uparrow t^+} \int_{t^-}^{t^+} (\bar{\mathbf{M}}\ddot{\mathbf{q}} + \mathbf{g}_{,\mathbf{q}}^T \boldsymbol{\lambda}) dt = 0. \tag{2.14}$$

The reduced force vector $\bar{\mathbf{f}}$ only contains non-impulsive forces and therefore the right-hand side vanishes. Under the introduction of the contact impulses,

$$\boldsymbol{\rho} = \lim_{t^- \uparrow t^+} \int_{t^-}^{t^+} \boldsymbol{\lambda} dt, \tag{2.15}$$

and noting that the mass matrix (in general a function of the generalized coordinates) remains constant during impact, the momentum equations for the system become

$$\bar{\mathbf{M}}\dot{\mathbf{q}}^+ + \mathbf{g}_{,\mathbf{q}}^T \boldsymbol{\rho} = \bar{\mathbf{M}}\dot{\mathbf{q}}^- \tag{2.16}$$

with $\dot{\mathbf{q}}^-$ the velocities before and $\dot{\mathbf{q}}^+$ the velocities of the system after impact. Together with Newton's impact law,

$$\dot{\mathbf{g}}^+ = -e\dot{\mathbf{g}}^-, \quad \text{or} \quad \mathbf{g}_{,\mathbf{q}}\dot{\mathbf{q}}^+ = -e\mathbf{g}_{,\mathbf{q}}\dot{\mathbf{q}}^-, \tag{2.17}$$

we have a complete set of linear equations reading

$$\begin{bmatrix} \bar{\mathbf{M}} & \mathbf{g}_{,\mathbf{q}}^T \\ \mathbf{g}_{,\mathbf{q}} & \mathbf{0} \end{bmatrix} \begin{bmatrix} \dot{\mathbf{q}}^+ \\ \boldsymbol{\rho} \end{bmatrix} = \begin{bmatrix} \bar{\mathbf{M}}\dot{\mathbf{q}}^- \\ -e\mathbf{g}_{,\mathbf{q}}\dot{\mathbf{q}}^- \end{bmatrix} \tag{2.18}$$

From these equations the velocities after impact $\dot{\mathbf{q}}^+$ together with the contact impulses $\boldsymbol{\rho}$ can be found. Because Newton's impact law (2.17) is often contradicted experimentally in case of multiple impacts, a restriction to simple impacts is made. The contact configuration for the walker is denoted by $u = 0$, $v =$ constant, and $\phi = 2\theta$. The velocities of the stance foot before impact are zero. Solving the impact equations at the contact configuration and subsequently taking the limit case yields for the velocities after impact

$$\lim_{m_f \to 0} \dot{\mathbf{q}}^+ = \begin{bmatrix} \dot{u}^+ \\ \dot{v}^+ \\ \dot{\theta}^+ \\ \dot{\phi}^+ \end{bmatrix} = \begin{bmatrix} -\sin(\theta)\cos(2\theta)\sin(2\theta) \\ \cos(\theta)\cos(2\theta)\sin(2\theta) \\ \cos(2\theta) \\ \cos(2\theta)(\cos(2\theta) - 1) \end{bmatrix} \dot{\theta}^-, \tag{2.19}$$

and for the contact impulses

$$\lim_{m_f \to 0} \boldsymbol{\rho} = \begin{bmatrix} \rho_x \\ \rho_y \end{bmatrix} = \begin{bmatrix} -\sin(\theta)\sin(2\theta) \\ \cos(\theta)\sin(2\theta) \end{bmatrix} \dot{\theta}^-. \tag{2.20}$$

The model is so simple that the resulting equations can be verified easily. First, the stance foot velocity after impact (Eq. 2.19) is $\cos(2\theta)\sin(2\theta)\dot{\theta}^-$ in the direction of the stance leg, this is the hip velocity after impact projected on this leg. Second, the contact impulse at the heel strike (Eq. 2.20) is directed along the swing leg with magnitude $\sin(2\theta)\dot{\theta}^-$, which is the projection of the hip velocity just before impact on the swing leg. Last, the velocities after impact (Eq. 2.19) are only a function of the stance leg angle θ and its angular velocity $\dot{\theta}^-$. This velocity is in fact the hip velocity. As a result, if we start the walker with the initial conditions on the state as $(\theta, \phi, \dot{\theta}, \dot{\phi})_0$, then after the first heelstrike (2.19) two initial conditions drop out and the next state is only dependent on θ and $\dot{\theta}^-$.

2.4.4 Stride Function

The mapping from the initial conditions $\mathbf{v} = (\mathbf{q}, \dot{\mathbf{q}})$, from one step to the next is what McGeer calls the 'stride function' [110], reading

$$\mathbf{v}_{n+1} = \mathbf{S}(\mathbf{v}_n). \tag{2.21}$$

We look for a motion of the walker were the legs return to the same state after one heelstrike, the so called period-one gait cycle. For the analysis of the gait we have to swap the stance and swing leg variables from step n to step $n+1$ as

$$\begin{aligned} \theta_{n+1} &= \theta_n - \phi_n \\ \phi_{n+1} &= -\phi_n. \end{aligned} \tag{2.22}$$

At heelstrike, the swing leg angle ϕ^- is equal to $2\theta^-$, and combining the time derivatives of (2.22) with the velocities after impact (2.19), gives us the initial conditions after heelstrike as

$$\begin{aligned} \theta_{n+1} &= -\theta_n^- \\ \phi_{n+1} &= -2\theta_n^- \\ \dot{\theta}_{n+1} &= \cos(2\theta^-)\dot{\theta}_n^- \\ \dot{\phi}_{n+1} &= \cos(2\theta^-)(1 - \cos(2\theta^-))\dot{\theta}_n^-. \end{aligned} \tag{2.23}$$

The stride function for the simplest walker is now complete. It starts with $(\theta_n, \dot{\theta}_n)$ as the initial conditions at the beginning of the nth step, which are numerically integrated with the equations of motion (Eq. 2.5) until heelstrike occurs. Then the velocities after heelstrike are calculated and the legs are swapped (Eq. 2.23), resulting in the initial conditions $(\theta_{n+1}, \dot{\theta}_{n+1})$ of the next step.

2.4.5 Cyclic Motion

Usually, the stride function has two cyclic solutions: initial conditions that map onto themselves. If started with those initial conditions, the walker shows a perfect repetitive motion. That motion is called a limit cycle, and the corresponding initial conditions are called the fixed point. The method for finding cyclic gait, as commonly used in passive dynamic walking research, is as follows. A walking cycle is specified by the requirement that the vector of initial conditions $\mathbf{v}_n$ results in identical initial conditions for the kth subsequent step:

$$\mathbf{v}_{n+k} = \mathbf{v}_n \tag{2.24}$$

A vector with initial conditions satisfying this requirement is a cyclic solution $\mathbf{v}_c$, which maps onto itself:

$$\mathbf{S}^k(\mathbf{v}_c) = \mathbf{v}_c \tag{2.25}$$

The main interest is symmetric walking ($k = 1$), i.e. a period-one gait cycle. Such cyclic solution can be found by a linearization of the stride function

$$\begin{aligned} \mathbf{S}(\mathbf{v} + \Delta\mathbf{v}) &\approx \mathbf{S}(\mathbf{v}) + \mathbf{J}\Delta\mathbf{v} \\ \text{with} \quad \mathbf{J} &= \tfrac{\partial \mathbf{S}}{\partial \mathbf{v}} \end{aligned} \tag{2.26}$$

and applying a Newton-Raphson iteration procedure, starting with a set of initial conditions $\mathbf{v}$ close to the cyclic solution $\mathbf{v}_c$

$$\begin{aligned} &\texttt{repeat} \\ &\quad \Delta\mathbf{v} = [\mathbf{I} - \mathbf{J}]^{-1}(\mathbf{S}(\mathbf{v}) - \mathbf{v}) \\ &\quad \mathbf{v} = \mathbf{v} + \Delta\mathbf{v} \\ &\texttt{until } |\Delta\mathbf{v}| < \epsilon \end{aligned} \tag{2.27}$$

where $\mathbf{I}$ is the identity matrix. The Jacobian $\mathbf{J}$ is calculated by a perturbation method, which involves simulation of a full walking step for every initial condition. The eigenvalues of $\mathbf{J}$ quantify the stability of the cyclic motion. If both eigenvalues are inside the unit circle in the complex plane, the limit cycle is stable and the robot can recover from small disturbances.

2.4.6 Results

Here we present the results exactly as they were presented by Garcia *et al.* [54]. In our experience, it is extremely useful to check your own simulations by finding an exact match with these results.

A typical plot of the leg angles over one step is shown in Fig. 2.6. For this plot, we used a slope angle of $\gamma = 0.009$ rad. Stable walking cycles can be found for slopes of $0 < \gamma < 0.0151$ rad. For larger slopes, we can still find limit cycles, but they are not stable. This is shown in Fig. 2.7 with the upper line. This figure also shows the existence of a second limit cycle, the lower line. To distinguish the two: the upper line is called the 'long period gait', while the lower line is the 'short period gait'. In our experience, almost all passive-based walking models yield these two solutions, and in almost all cases the 'short period gait' is unstable. Therefore, we usually focus on only the 'long period gait'. For this gait, the eigenvalues of the stride function are plotted in Fig. 2.8. At $\gamma = 0.151$, one of the eigenvalues goes through -1, meaning that the gait becomes unstable.

We have to devote a few words on the difference between finding a periodic cycle and finding a stable solution. A cycle is found by looking for initial conditions that result in themselves after one step, using Eq. 2.27. Such a solution can be stable or unstable. The stability can only be determined *after* the cycle has been identified. Of course, one could look for periodic solutions by just running a forward simulation and waiting many steps. If a cycle exists *and it is stable*, and the search has been started sufficiently close to the cycle, then one can find it. This is analogous to just starting a prototype over and over again, hoping that it will walk. However, this trial-and-error method will not indicate how to change an unstable design in order to make it stable. Therefore, we strongly advocate the two-tiered approach; first find a cycle with Eq. 2.27, and then determine its stability with the eigenvalues of **J** from Eq. 2.26. If a cycle is found but it happens to be unstable, then one can investigate the effect of parameter changes on the stability and thus improve the model until a stable solution is found. And in some cases, we might not even care about stability; perhaps we intend to just find the most efficient motion, which will then be stabilized with a controller to be designed around that efficient passive motion.

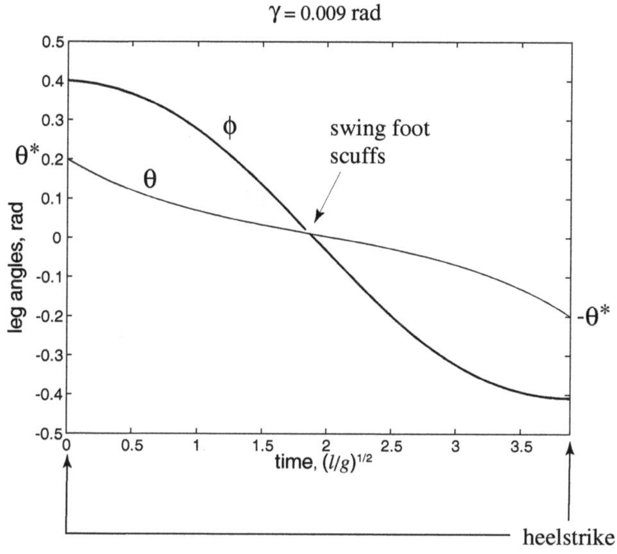

Fig. 2.6. Typical step. Following fig. 2.5, θ is the absolute stance leg angle and ϕ is the relative swing leg angle. The step starts with the fixed point values, such as θ^*. Reprinted with permission from Garcia *et al.* [54].

2.5 Practical Hints and Tips

Over the years, we have received many requests for practical advice and tips for building physical prototypes of passive dynamic walkers. The first advice is to go through the exercise of simulating the Simplest Walking Model above. Even if you don't intend to make a realistic simulation model of your new prototype, the simple model will help you to get a feeling for the dynamics and for the various ways that the walker can fall. The second advice is perhaps somewhat

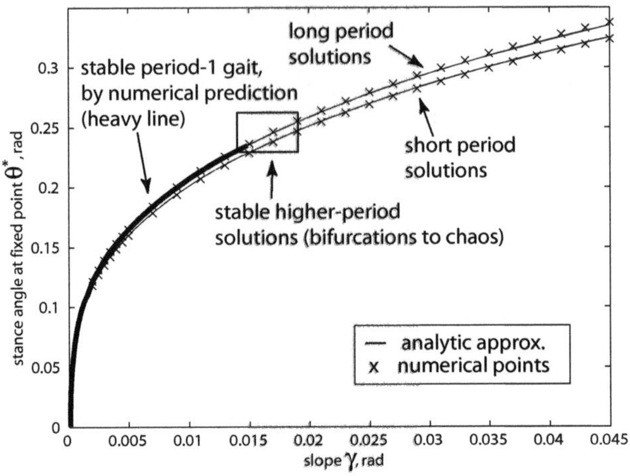

Fig. 2.7. Stance leg angle θ at fixed point versus slope angle γ. Reprinted with permission from Garcia *et al.* [54].

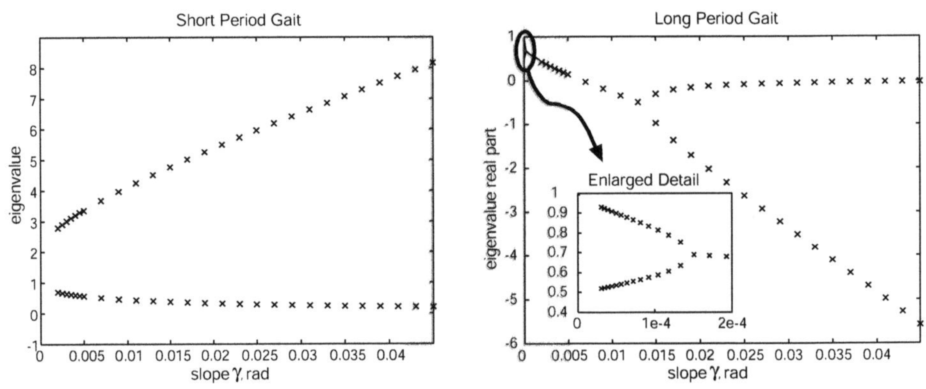

Fig. 2.8. Eigenvalues of short-period gait (left) and long-period gait (right) as a function of the slope angle γ. The gait is stable if all eigenvalues are between -1 and 1. Reprinted with permission from Garcia *et al.* [54].

unexpected. Instead of building a fully passive machine, it is probably much easier to build a machine with actuation in the hip joint. In Chapter 4, we explain how this will greatly improve the disturbance handling of the robot. Our experience is that robots with some actuation at the hip joint are far easier to tune and to launch than the fully passive ones.

Other than that, we have the following points of advice:

- **Arc feet.** Do not use point feet because the basin of attraction is very small [141]. Arc feet are better, with a radius of approximately 1/3 of the leg length [108, 186]. The arc feet are rigidly attached to the shank, so there is no ankle joint. If you don't like this non-human feature, the arc feet could equally well be replaced with flat feet that are mounted on ankles with a spring in it [179].

- **Fore-aft mass balance.** For each new slope angle, you will probably need to re-adjust the fore-aft mass balance. There are two effects to be considered simultaneously. First, the fore-aft mass balance affects the amount of acceleration and deceleration within each step. When the robot has a tendency to slow down and fall backward, place the mass more forward. If you overdo this, the robot will accelerate and fall forward. The mass balance here is in strong relation with the position of the foot. If the foot is displaced forward (as in most kneed passive walkers [109]), then the mass also needs a forward offset. Second, the mass balance within a leg affects the passive swing motion. The swing leg will move symmetrically around its vertical equilibrium. So if the mass is placed a bit backward, then the leg swings out further than it started. This is good, because it decreases the chances of a forward fall [184]. As McGeer already noted; friction in the hip joint can be countered by placing the leg mass a bit more backward.
- **Faking 2D.** Most researchers start with the 2D passive walkers, using the four-legged symmetric construction (e.g. Fig. 2.3). The robot is not really 2D (nothing is, in our physical world), and so it is important to make the approximation as good as possible. Good solid engineering work is required, as is a nice floor surface. An additional trick is to couple the outer shanks of a kneed walker [82].
- **Good decent engineering** is so important that we mention it again. Bouncing, flexibility, friction, and joint play (all of which cannot accurately be modeled) should be reduced to a minimum.
- **Mass high up.** In general, the mass should be high up in the legs; heavy hips are good, heavy feet are bad. This is also true for a model with knees; the shank should be lightweight, and its center of mass should be close to the knee joint.
- **No knee collapse.** The knee is prone to collapsing during the stance phase. It is difficult to construct a passive solution to this problem, because there are two causes. First, the ground reaction force during stance may be such that it will cause the knee to buckle. The solution for this problem is forward-placed feet as pointed out by McGeer. Second, the knee may not be fully extended at heel strike. This problem is caused by bouncing during the knee strike. This is quite a persistent problem, due to flexibility in the hyperextension stop and also due to flexibility of the leg segments themselves. Note that using "dead rubber" in the hyperextension stop will not work; the flexibility of the leg will still store energy which will still release in the form of a bouncing knee. After experimentation with mechanical catches, McGeer finally settled on using leaky suction cups. These suction cups hold the knee in extension just after knee strike, but thanks to the leak they will release by the time that the next swing phase starts. The amount of leakage has to be tuned carefully. For our robots, we have settled on using mechanical latches that

are actively released at the start of the swing phase. And for our new robots which will have some actuation anyway, we will use a torque that drives the knee into its hyperextension stop, preventing knee collapse during stance.

- **Symmetric gait.** The legs should be equal so that the robot is symmetric. If the result is still an asymmetric gait (a long step followed by a short step), then this means that the walking cycle is just barely stable. In the simplest walking model [54], this occurs at steep slopes. Try to adjust the parameters such that the asymmetry goes away.
- **Hip end stops.** The stability will benefit from limit stops at the hip joint [184]. The result is a more constant step length. At each step, the swing leg should bump into the hip limit stop, and then (slightly) fall back [177].

3. Baps; Pneumatic Actuation

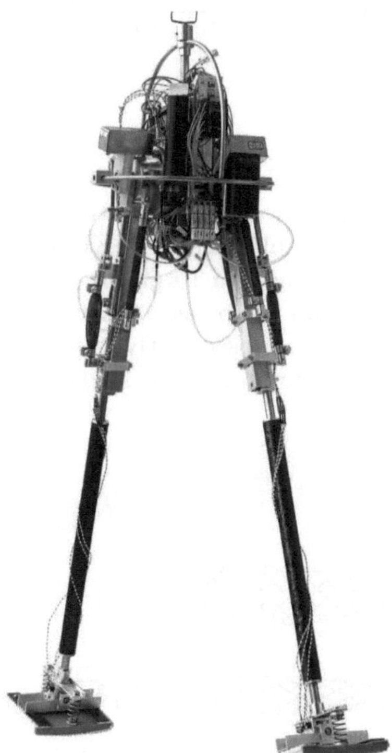

M. Wisse and R.Q. van der Linde: Delft Pneumatic Bipeds, STAR 34, pp. 25–44, 2007.
springerlink.com

This chapter reports the development of the pneumatic components that we used for all robots in this book. The choice for pneumatic actuation has advantages as well as drawbacks compared to actuation with electric DC motors. The advantage is the inherent compliance of the McKibben muscles (Section 3.1). The good match between the compliant McKibben muscles and the concept of passive dynamic walking was the main argument in favor of the choice for pneumatics. The main drawback is the lack of sufficiently lightweight commercially available components, which means that many of the components must be specially developed for the research project (Section 3.2). A second drawback is the difficulty of implementing high-bandwidth control, which may be a reason for us to switch to electric DC motors for our future prototypes. However, for the phasic (i.e. once-per-step) actuation that we have implemented in all robots in this book, the pneumatic system is a satisfactory choice (Section 3.3). The complete, autonomous pneumatic system can successfully power a biped robot as demonstrated with the prototype 'Baps' (Section 3.4). The content of this chapter is based on four of our papers [157, 159, 158, 186].

3.1 McKibben Muscles

A McKibben muscle is a flexible rubber tube, covered by a weave of flexible yet non-extensible threads, see Fig. 3.1. One end of the tube is closed, the other is the pneumatic inlet. With this simple design, McKibben muscles are cheap, easy to attach, compliant, and lightweight (a 10 gram muscle can pull 40 N). These characteristics are the reason for choosing McKibben muscles instead of hydraulic or electric actuators. And, subsequently, this choice underlies the choice for pneumatics and thus forms the fundament for the entire book.

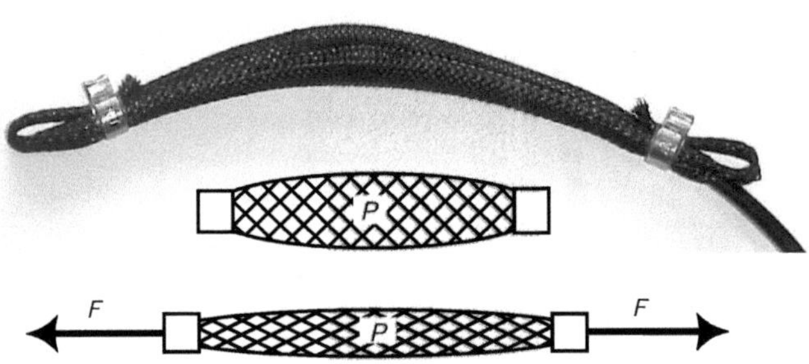

Fig. 3.1. McKibben muscle. **Top:** Commercially available McKibben muscle from Shadow [6]. **Bottom:** Muscle operating principle.

The compliance of the McKibben muscles is key. The resulting resemblance with human muscles was the reason for their initial development by Joseph L. McKibben in the 1950s. The initial field of application was rehabilitation [56, 138, 121, 49, 55, 69, 174, 154, 59, 51], later to be extended to robotics [84, 67, 89, 20, 19, 21, 125, 156, 152, 151, 27, 94, 161, 71, 126, 101, 28, 29, 9, 44, 72, 153, 93,

90, 91, 92, 43, 42, 63, 11, 13, 155, 37, 14, 137, 22, 146, 145, 147, 144, 77]. Because of their compliance, the muscles are particularly successful in applications that do not require a high bandwidth nor a high position accuracy. This fits well with the original concept of passive walking; the passive hip joint can be regarded as an extreme form of compliant actuation.

The force-length relation can be theoretically derived from the geometry of the muscle structure. Here we follow the derivation as originally published by Chou and Hannaford [26]. When the muscle length changes due to an external force F, its volume also changes (Fig. 3.1). For this change of volume dV at the relative pressure difference with atmosphere P', the following amount of virtual internal work dW_{int} is required:

$$dW_{int} = P'dV \tag{3.1}$$

This virtual internal work is in balance with the virtual external work dW_{ext} done by the muscle on the robot system

$$dW_{ext} = -FdL \tag{3.2}$$

where F is the external force and dL the virtual elongation. Combining Eq. (3.1) and (3.2) results in

$$F = -P'\frac{dV}{dL} \tag{3.3}$$

We now make the following two assumptions:

- The muscle shape is modeled as a perfect cylinder (no deformations at the end caps).
- The threads of the braiding are infinitely stiff in longitudinal direction.

The relation between the elongation dL and the volume change dV can then be found using the thread length b, the number of thread windings n, the angle θ between the threads and the cylinder axis, and the diameter D of the cylinder:

$$L = b\cos\theta \tag{3.4}$$

$$D = \frac{b}{n\pi}\sin\theta \tag{3.5}$$

$$V = \frac{1}{4}\pi D^2 L = \frac{b^3}{4\pi n^2}\sin^2\theta\cos\theta \tag{3.6}$$

Substituting this relation in Eq. (3.3), we obtain

$$F(\theta) = \frac{b^2 P'}{4\pi n^2}(3\cos^2\theta - 1) \tag{3.7}$$

or

$$F(L) = \frac{b^2 P'}{4\pi n^2}(3\frac{L^2}{b^2} - 1) \tag{3.8}$$

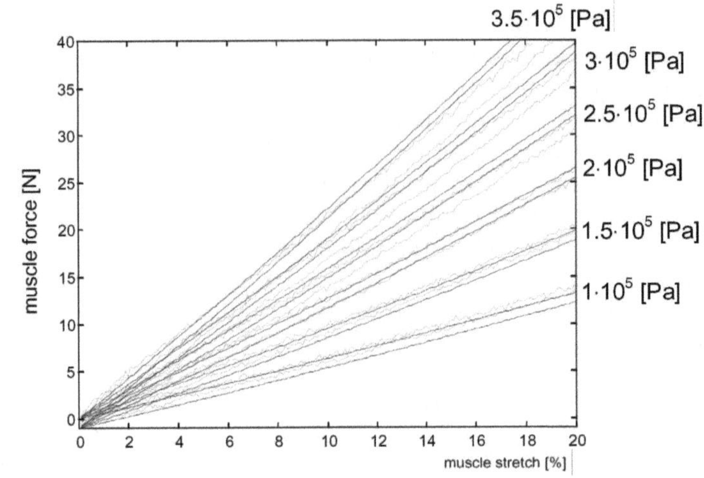

Fig. 3.2. Force-length relation for the linear range of the McKibben muscles. The dark lines are created with the muscle model in eq. (3.8), with parameters according Table 3.2. The grey lines are quasistatic measurements. The graph shows that the model provides an accurate prediction, and that the muscles behave nicely linear.

Up to 20% extension, the force-length relation is approximately linear. Fig. 3.2 shows the relation for pressures between 0.1 MPa and 0.35 MPa. The figure clearly shows good correspondence between the theoretical model and the experimental measurements. A small amount of unmodeled hysteresis is present, due to friction between the rubber tube and the braiding. Within the linear range, the muscles can be regarded as springs with actively variable passive stiffness. The word 'actively' stands for the fact that the stiffness can be changed by adjusting the pressure, whereas 'passive' stands for the fact that no further power (no pneumatic flow) is required for the spring-like behavior. In the prototype Baps, the McKibben muscles are used in the linear range with pressures between 0.17 MPa and 0.25 MPa.

Beyond 20% extension, two non-linear effects come into play. First, the nonlinear terms in the force-length relation Eq. 3.8 become significant, leading to a gradual stiffness increase. Second, the muscles reach their maximal extension at approximately 30%. At this point, the braiding threads are adjacent, i.e. they touch each other and have no room for further motion. Most researchers will avoid the nonlinearity and especially the region of maximal extension, but we have made good use of it for our robots Mike (Chapter 4), Max (Chapter 5), and Denise (Chapter 6). In the hip joints of these robots, the muscles are not only used as actuators, but also as joint angle limits. Chapter 4 explains why such a hip joint limit is beneficial for stability. The nonlinear force-length relation is plotted in Fig. 3.3. When modeling this behavior for computer simulations, we use a second linear model with different parameters for the region between 20% and 30% elongation. To determine the parameter values for this part of the

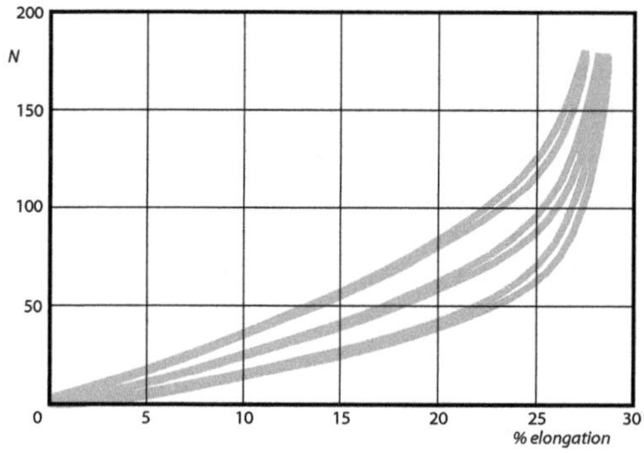

Fig. 3.3. Measured muscle force-length relation at three different pressures. Above 20% elongation the behavior becomes strongly non-linear. The maximal extension is approximately 30%.

muscle model (including a particularly high damping near maximal extension), our practice is to match the walking motions of the full model to the measured walking motions, rather than fitting the linear approximation directly to the measurements in Fig. 3.3.

McKibben muscles are fairly easy to build. We built our own muscles for the prototype Baps, in order to optimize the material choice for minimal friction losses. Table 3.1 shows various combinations of tubing and braiding material. We measured the spring efficiency for these material combinations by performing quasi-static tests. For a given pressure, the muscle was slowly extended and then slowly returned back to the unloaded length, resulting in force-length graphs such as Fig. 3.2. The values in Table 3.1 are obtained by dividing the amount of work lost to friction (the area in the hysteresis loops in Fig. 3.2) by the total amount of energy that was stored at 20% extension. The table shows that a combination of polyester braiding and latex tubing resulted in the highest efficiency, and therefore this combination was selected for the muscles of Baps. In hindsight, we have to report that the latex tubing deteriorates in a few months, and that it was difficult to produce sufficiently strong end caps for (ab)using the muscles as extension stops.

Table 3.1. Maximum energetic efficiencies from isobar, quasi-static measurements of different material combinations for the McKibben muscle. Latex tubing has a wall thickness of 0.2 mm, and silicone tubing has a wall thickness of 1 mm.

Braiding material	Latex tubing 0.2 mm	Silicone tubing 1 mm
Polyester	0.94	0.87
Nylon	0.87	0.79
Carbon fiber	0.73	0.69
Kevlar®	-	0.69

For the later prototypes (Chapters 4 to 6), we decided not to build our own muscles, but to use commercially available muscles. Those produced by Shadow company [6] best suited our size and weight requirements (10g muscles for 40N). For some time in the past, Bridgestone was selling McKibben muscles in Japan under the tradename 'Rubbertuator' [84, 39, 40]. Currently, the German company Festo sells air muscles (used by [102, 15, 14, 90, 91, 92]) and the Japanese Hitachi Medical Corporation sells rubbertuators in Japan (used by [146, 145, 147, 144, 77]).

3.2 Pressure Control Unit

3.2.1 Background and Requirements

The pressure control unit must be able to regulate the desired muscle pressures accurately and fast (i.e., well within the step time of 0.6 seconds). In addition, the application in an autonomous biped requires a compact, lightweight and gas efficient solution.

There are two commercially available regulator principles, each of which can only fulfill part of the above requirements. The first type is the flapper-nozzle type. These indirectly controlled pressure regulators provide fast and accurate pressure control at the cost of a high internal gas consumption and relatively large physical dimensions. The second type is the piston type. These directly controlled pressure regulators are generally small and lightweight and need no extra gas supply for internal consumption, but they are not sufficiently sensitive and accurate for our application. Nevertheless, we decided to use the piston type, as small size and gas efficiency are the most important requirements. We minimized the disadvantages of this type by developing our own design.

3.2.2 Operating Principle

The piston type pressure regulator is drawn in Figure 3.4. A valve separates the input pressure from the output pressure. The output pressure acts on a spring loaded piston, where the manually adjustable spring load represents the output pressure level. If the output pressure falls below this desired value, the spring loaded piston opens the valve and the output pressure level is restored. To ascertain that the pressure regulator is sensitive, it needs to be constructed with a high ratio of A:C (see Figure 3.4) and with low internal friction.

The low preset pressure level is realized by connecting a separate back-pressure regulator (see Figure 3.4) via a controlled valve to the muscle outlet. The spring loaded piston in the back-pressure regulator is open as long as the muscle pressure is higher than the desired level (drawn situation).

3.2.3 Technical Realization and Results

The principles discussed above have been translated into functioning prototypes. Experiments have convinced us that the required high accuracy (10 kPa) cannot be

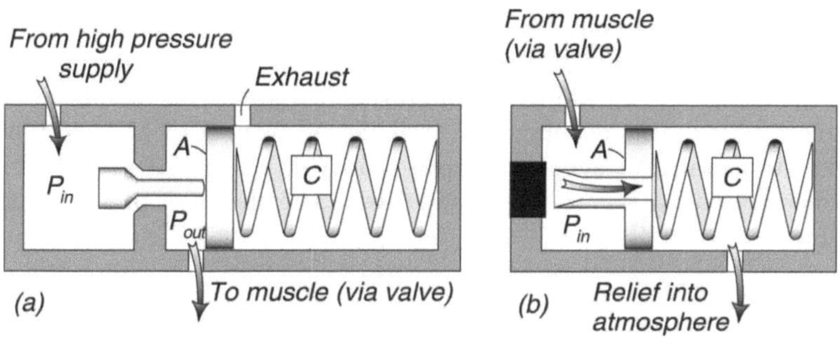

Fig. 3.4. Working principle of (a) pressure regulator and (b) back-pressure regulator

met by a single-stage pressure regulator, due to pressure overshoot and hysteresis. Therefore we have divided the pressure reduction in two stages, see Figure 3.5.

First, one main pressure regulator directly on the gas canister brings the pressure from 5.8 MPa to about 1.0 MPa. Second, a second-stage reduction from 1.0 MPa to 0.2 – 0.4 MPa, with 4 different preset manually adjustable pressure levels, is realized in the input pressure control block ('IN', Figure 3.5). In these valves, the pistons are equipped with diaphragms to minimize friction effects and to provide the required sensitivity and accuracy. The output pressure control block ('OUT') includes four adjustable back-pressure regulators. Basically the same piston construction as in the input pressure reduction valves has been used. The two pressure-control blocks together weigh about 180 gram and have a volume of less than 8 x 5.5 x 1.5 cubic centimeter.

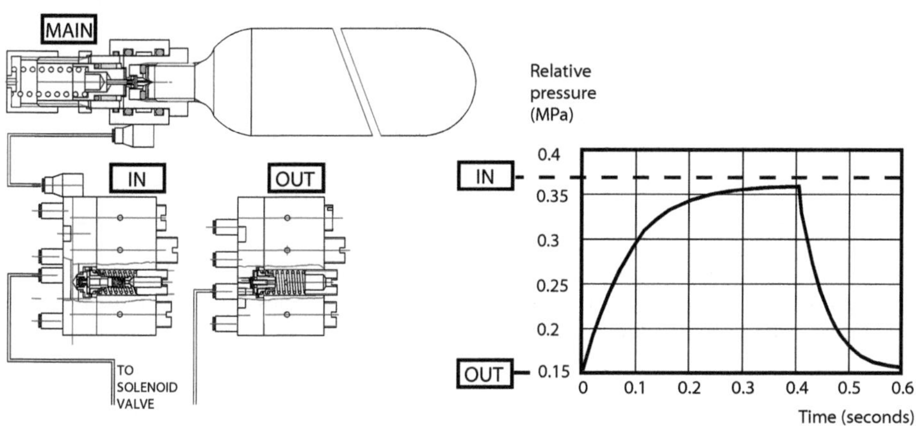

Fig. 3.5. Left: technical drawings of pressure reduction system. **Right:** dynamic response of the complete pneumatic system.

After assembling the complete pneumatic system, it is possible to evaluate the behavior by measuring the muscle pressure in time, during a switching-action of the described solenoid valve. Figure 3.5 shows the dynamic response of the complete pneumatic system when pressurized from 0.15 MPa to 0.35 MPa and back. We obtain an accuracy/repeatability of about 10 kPa, and a relatively slow response as was to be expected with the choice of pressure regulator type.

3.3 Phasic Activation of a Single Pendulum Test Setup

The next step in the research was to test the functioning of the complete pneumatic actuation system in an oscillating periodic motion. We chose a single pendulum as a test setup. This system has a well-defined natural frequency and can easily be modeled. It can be regarded as a model for swing leg behavior during walking. Two artificial pneumatic McKibben muscles were used as an agonistantagonist couple (Fig. 3.6).

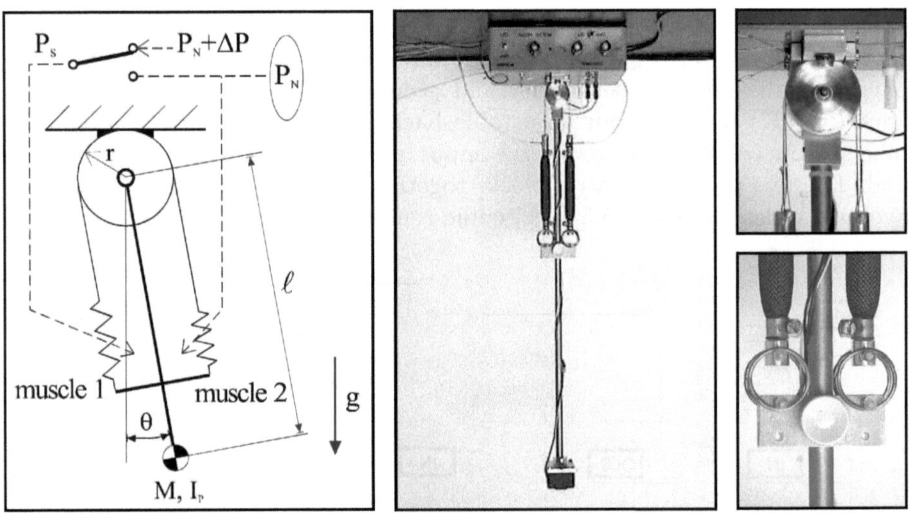

Fig. 3.6. Pendulum test setup. **Left:** Model and pressure schema. **Middle:** Overview of complete setup. **Right:** Friction-less attachment of muscles at fixed moment arm r. This figure originally appeared in [158], © 1999 IEEE.

The system can be brought into a limit cycle by applying periodic activation. This activation should not be a function of time (such as a pulse each second, which would result in a *forced oscillation*) but it should be triggered by the motion itself so that an autonomous periodic motion emerges. One can choose any point in the cycle. We have chosen to use the maximal excursion of the joint angle as trigger, which is equal to a downwards triggered zero-crossing of the angular rate. Therefore a gyroscope was mounted at the end of the pendulum. The activation is obtained by raising the pressure by ΔP in muscle 1. In the set-up a three-way valve was used to switch the pressure in muscle 1 between the nominal pressure and the activation pressure (Fig. 3.6). When an activation time Δt has elapsed the supply pressure P_S is set back to the nominal pressure P_N again. The switching time of the valve is very small compared with the time scale of an activation pulse, and will therefore be neglected. Muscle 2 is kept at a constant pressure P_N.

3.3.1 System Model

The pneumatic muscle dynamics can be simplified by a lumped parameter model. The muscle volume (V_M) can be modeled as a constant accumulator, or capacitor (C_M). The hose can be modeled as a resistor (R). Modeling the activation muscle pressure (P_{AM}) dynamics then results in a first order filter, given by the differential equation

$$\dot{P}_{AM}(t) + \frac{P_{AM}(t)}{C_M R} = \frac{P_S(t)}{C_M R} \tag{3.9}$$

where C_M can be expressed by

$$C_M = \frac{\Phi_{AM}}{\dot{P}_{AM}} = \frac{dm/dt}{dP_{AM}/dt} = \frac{dm}{dP_{AM}} \tag{3.10}$$

where Φ_{AM} is the mass flow through the hose and m is the gas mass. Using the ideal gas equation, Eq. (3.10) can be rewritten as

$$C_M = dm \cdot d(\frac{V_M}{mR_{GAS}T}) \approx \frac{V_M}{R_{GAS}T} \tag{3.11}$$

where R_{GAS} is the specific gas constant and T is the gas temperature.

We model the dissipating friction forces as a viscous damping. This results in the following equation of motion for the complete pendulum setup of Fig. 3.6:

$$I_P\ddot{\theta}(t) = -Mgl\sin\theta(t) - F_{M1}(\theta(t), P_1(t))r + F_{M2}(\theta(t), P_2)r - 2kr^2\dot{\theta}(t) \tag{3.12}$$

with

- I_P pendulum inertia around the joint;
- M pendulum mass;
- g gravitational acceleration;
- l distance from joint to center of mass of pendulum;
- F_{Mn} force of muscle n as a function of the pendulum angle θ and muscle pressure P_n;
- r radius of the cylinder;
- k constant of damping of one muscle.

The muscle forces in this equation, F_{M1} and F_{M2}, can be obtained by substituting Eq. (3.8). Combining the result with Eq. (3.12) yields the final expression of the complete pendulum dynamics:

$$\begin{aligned}\ddot{\theta} = & -\tfrac{Mgl}{I_p}\sin\theta - \tfrac{2kr^2}{I_p}\dot{\theta} \\ & -\tfrac{rb^2 P_{AM}(t)}{4\pi n^2 I_p}\left(\tfrac{(U_0+r\theta)^2}{L_0^2} + \tfrac{2(U_0+r\theta)}{L_0}\right) \\ & +\tfrac{rb^2 P_N(t)}{4\pi n^2 I_p}\left(\tfrac{(U_0-r\theta)^2}{L_0^2} + \tfrac{2(U_0-r\theta)}{L_0}\right)\end{aligned} \tag{3.13}$$

Table 3.2. System parameters of the pendulum test setup of Fig 3.6

β	0.026	U_0	0.01 m
C_M	$1.55 \cdot 10^{-10}$ kg · m²/N	g	9.81 m/s²
R	$2.80 \cdot 10^{8}$ N · s/m	l	0.273 m
R_{AIR}	287.00 J/kg · K	b	0.1126 m
r	0.018 m	m	0.309 kg
V_M	$13.1 \cdot 10^{6}$ m³	L_0	0.065 m
T	293 K	I_p	0.0317 kg · m²
n	1.83		

For the parameter values of the model, we used the values obtained from measurement of the pendulum setup, see Table 3.2. The damping coefficient β was determined by curve fitting of unactuated responses of the pendulum. This dimensionless damping ratio could be well approximated as a constant value for the frequencies in the range of interest. Its relation with the damping constant k used in Eq. (3.13) is given by:

$$k = \frac{\beta \omega_B I_p}{r^2} \tag{3.14}$$

where ω_B is the natural frequency of the pendulum. The time constant of the activation muscle can be calculated as $\tau_{AM} = RC_M = 0.043$s.

3.3.2 Model Validation and Parameter Sensitivity

We performed experiments with various settings for the nominal pressure, pressure difference, and activation time. Five measured fixed points were compared with the predicted fixed points. Table 3.3 gives the errors between the measured and predicted fixed points for the five settings. According to Table 3.3, the cycle amplitude of a fixed point cannot be predicted accurately. This can be clarified with a parameter sensitivity test. With the pendulum model the shift of the fixed point was verified with a 10% parameter change. The calculated parameter sensitivity matrix yields:

$$\begin{bmatrix} \frac{\partial\theta_F}{\partial\Delta t} & \frac{\partial\theta_F}{\partial\beta} & \frac{\partial\theta_F}{\partial\tau_{AM}} & \frac{\partial\theta_F}{\partial L_0} \\ \frac{\partial T_F}{\partial\Delta t} & \frac{\partial T_F}{\partial\beta} & \frac{\partial T_F}{\partial\tau_{AM}} & \frac{\partial T_F}{\partial L_0} \end{bmatrix} = \begin{bmatrix} 0.7838 & -5.1925 & -0.0394 & -1.6879 \\ 0.0784 & -0.3926 & 0.1453 & 3.6173 \end{bmatrix} \tag{3.15}$$

Table 3.3. Errors between model predictions and experimental measurements of the pendulum amplitude and period, for five different activation settings

measurement	Δt s	ΔP Pa	P_N Pa	amplitude error (%)	period error (%)
1	0.183	$0.15 \cdot 10^5$	$1.25 \cdot 10^5$	12	2.6
2	0.183	$0.15 \cdot 10^5$	$1.50 \cdot 10^5$	7.6	3.8
3	0.183	$0.20 \cdot 10^5$	$1.50 \cdot 10^5$	25	1.8
4	0.112	$0.40 \cdot 10^5$	$1.85 \cdot 10^5$	27	1.2
5	0.424	$0.10 \cdot 10^5$	$1.50 \cdot 10^5$	30	1.1

Eq. (3.15) shows that the cycle amplitude is highly sensitive to variations of the relative muscle damping β. Such variations of β are likely to occur due to temperature changes or ageing effects of the muscle. Also the pendulum model has a simple linear viscous damper, whereas real McKibben muscles will have a speed dependent constant of damping, causing the model to deviate from the measurements. The relative muscle damping is a critical design parameter, and is therefore assumed to cause the large deviations between predicted cycle amplitude and measured cycle amplitude in Table 3.3. Variation of the muscle zero-length L_0 has a large effect on the cycle amplitude and the cycle period. It is assumed to be constant in the muscle model of Eq. (3.8), but it varies for real McKibben muscles (Fig. 3.7). Variations in the muscle activation time Δt and activation muscle time constant τ_{AM} have relatively little effect on the location of the fixed point and can also be determined rather accurately.

From this result, we learn that the cycle amplitude is rather sensitive to the actual damping value. This is a logical result for any highly efficient system. A small amount of energy loss will have a great influence on the motion. We can predict that if amplitude is important (as it will determine the step length), then we will need to install an amplitude-regulating feature. In the Muse*on* walker

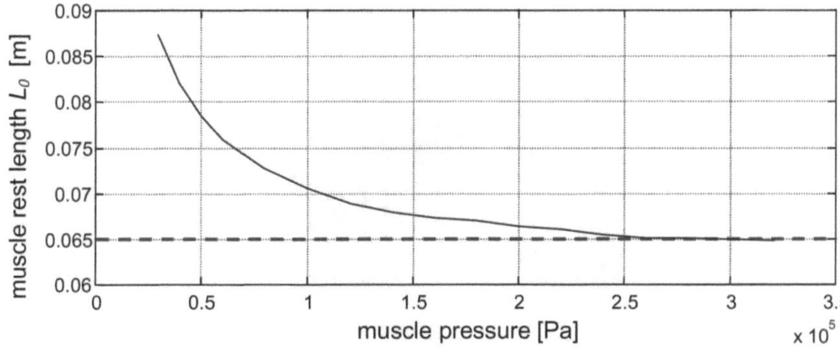

Fig. 3.7. Measured rest length of the applied McKibben muscle as a function of muscle pressure. The dashed line is the theoretical rest length, according to the muscle model of Eq. (3.8).

(Fig. 2.4), this was implemented by a limitation to the hip joint. In Chapter 4 such a feature will be analyzed in depth.

3.4 Prototype Design and Experiments

The previous sections have provided all the tools for pneumatic actuation. There is convincing evidence that actuation with McKibben muscles can successfully be combined with passive dynamic models to create active dynamic walking. The final test will be performed in this section, where we construct the autonomous biped 'Baps' and study its walking capabilities.

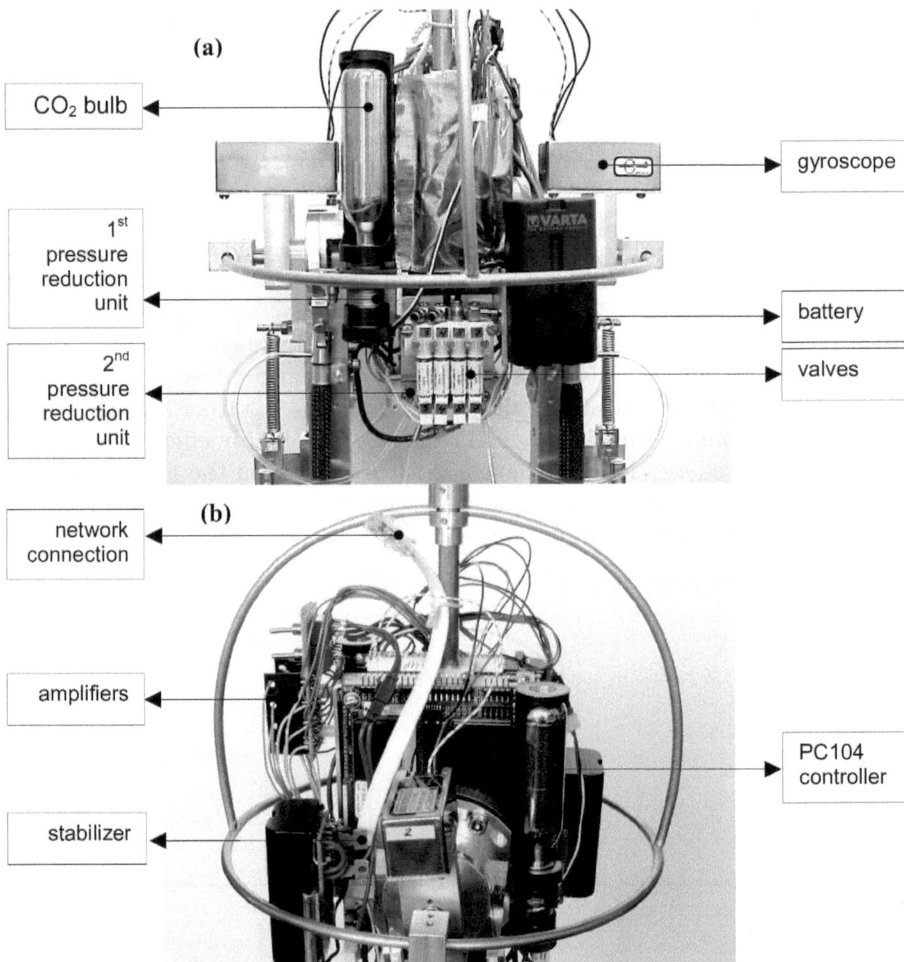

Fig. 3.8. Placement of electric and pneumatic components in the trunk

3.4.1 Specifications

The leg length is 65 cm, and due to the relatively small trunk the total length is about 80 cm. Baps weighs only 3.35 kg, subdivided as shown in Table 3.4. The pneumatic and electric components are mounted on the trunk of the biped, see Fig. 3.8. The location of the components is chosen such that the trunk center of gravity is located approximately in the middle between the hips. An outer cage construction of thin walled stainless steel tube is applied as a protection for the onboard equipment, and serves as a hand grip during experiments.

The location of the centers of gravity were determined with balancing tests. The inertia of the legs was derived from the measured frequency of free leg swing motions. The data is summarized in Fig. 3.9 and Table 3.5.

3.4.2 Muscle Configuration

Baps is equipped with six McKibben muscles, three per leg. Each leg has one muscle for leg elongation of a linear joint in the leg (push-off), and a pair of antagonistic muscles around the rotational hip joint, see Figs. 3.10-3.11. All muscles operate at a nominal pressure level providing a nominal stiffness, in addition to which there are short pressure increases at each step providing power to the motion.

3.4.3 Feet

Baps has had two types of feet; spherical feet and cylindrical feet, see Figs. 3.12 and 3.13. The first type was rigidly attached to the legs, whereas the second type was mounted on a spring-loaded ankle joint. We tested both.

We performed these tests because the feet have to fulfill four functions at once, and we could not predict which foot type would succeed best. First, the feet must allow lateral rocking. Therefore, we give the feet a radius in a frontal view, similar to the design of some ancient walking toys [172] and recent robot research [150].

Second, the feet must allow forward rotation of the stance leg. In McGeer's research, this was always done with a rigidly attached foot with a radius in the sagittal plane. For Baps, this means that there is a radius in two planes, and so the foot is effectively a part of a sphere. This is the first type of foot. Instead of an arc shape in the sagittal plane, one can also use a flat foot and a spring-loaded ankle joint. The resultant sagittal stance leg dynamics are similar [179] because of a similar forward travel of the center of pressure. Thus, the second type of feet for Baps has a spring-loaded ankle and a cylindrical shape with the curve in the frontal plane.

The third function that the feet must fulfill is to provide sufficient yaw resistance. Yaw, i.e. rotation around the vertical axis, is a source for instabilities which is induced by the counter-swinging motion of the two legs [36, 181]. Obviously, the spherical shaped feet will perform worse than the cylindrical feet,

Table 3.4. Weight, dimensions, and numbers of components in the autonomous biped Baps

function	component	mass [gram]	number	sizes [mm]
frame	leg	355	2	-
	trunk	1210	1	-
	foot (set A)	153	2	-
pneumatics	hip muscles	17	4	-
	leg muscles	17	2	-
	CO_2 holder	36	2	125x25x25
	CO_2 canister	59	2	88x∅22
	reduction unit	78	1	52x35x14
	valve unit	104	1	7x7x33
electronics	batteries	91	2	70x38x20
	PC	330	1	40x96x90
	gyroscope	56	3	58x25x25
total		3350		

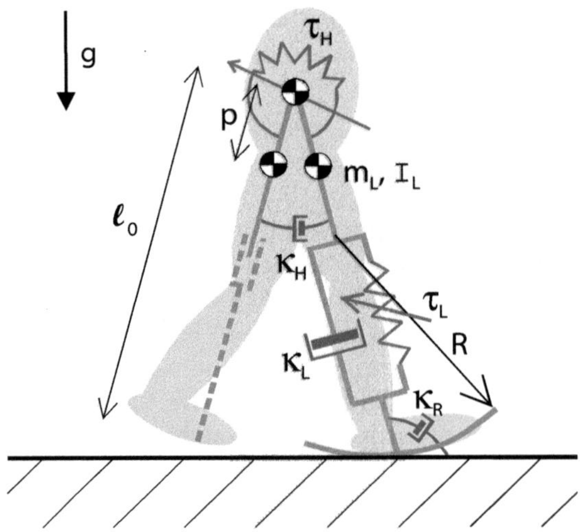

Fig. 3.9. Definition of the mechanical parameters of Baps

Table 3.5. Mechanical parameters of the autonomous biped Baps, according to Fig. 3.9

m_H	2.130 kg	I_H	0.0071 kgm^2
m_L	0.609 kg	I_L	0.0371 kgm^2
l_0	0.641 m	p	0.261 m
R	0.5 m	κ_R	0.751 Nms
τ_H	0.16 s	κ_H	0.0717 Nms
τ_L	0.10 s	κ_L	136 Ns/m

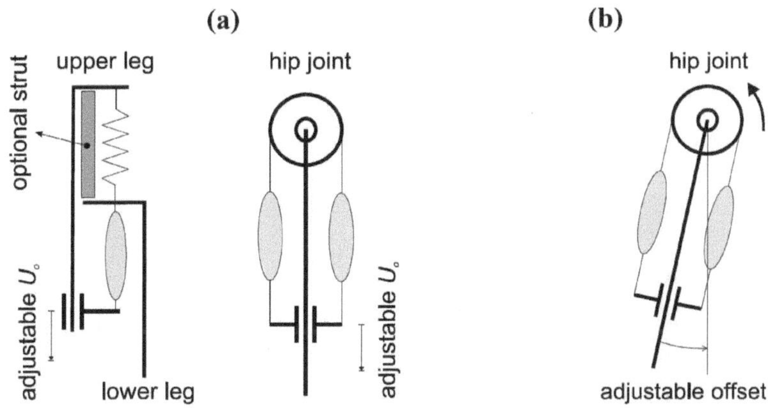

Fig. 3.10. Schematic of muscle attachments in Baps. (a) The pre-loaded length of the muscles in the legs and hips can be adjusted. (b) In the hips, the offset of the zero position can also be adjusted by rotating the cylindrical base.

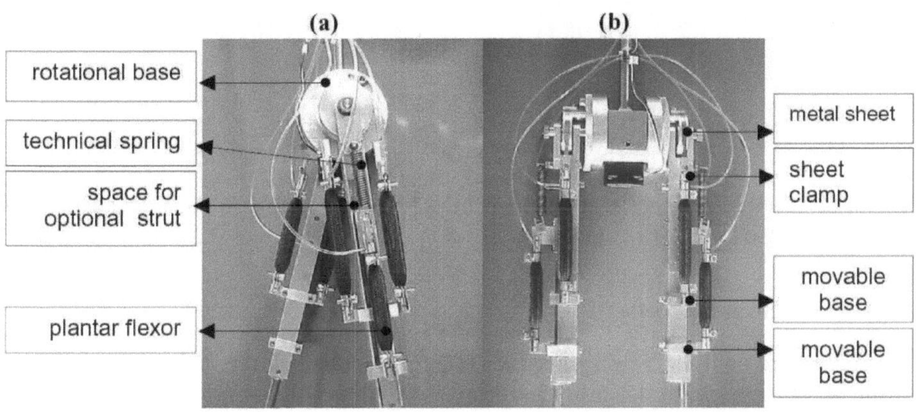

Fig. 3.11. A sagittal view (a) and a frontal view (b) of the assembled biped frame with the pneumatic muscles attached to it. Important components are indicated.

although the application of anti-slip rubber and the rather large radius (0.5 m) together perform reasonably well.

The fourth function is to provide sufficient foot clearance during the swing phase. For this function, the spherical feet are better. If we would have had actuation in the ankles or more clearance thanks to knee flexion, then the cylindrical feet would have sufficed too, so we expect that this foot design will be useful in the future. But in the prototype Baps, the toes and heels made inappropriate early contact with the floor. Thus, for this prototype we were not able to obtain successful walking motions with the cylindrical feet.

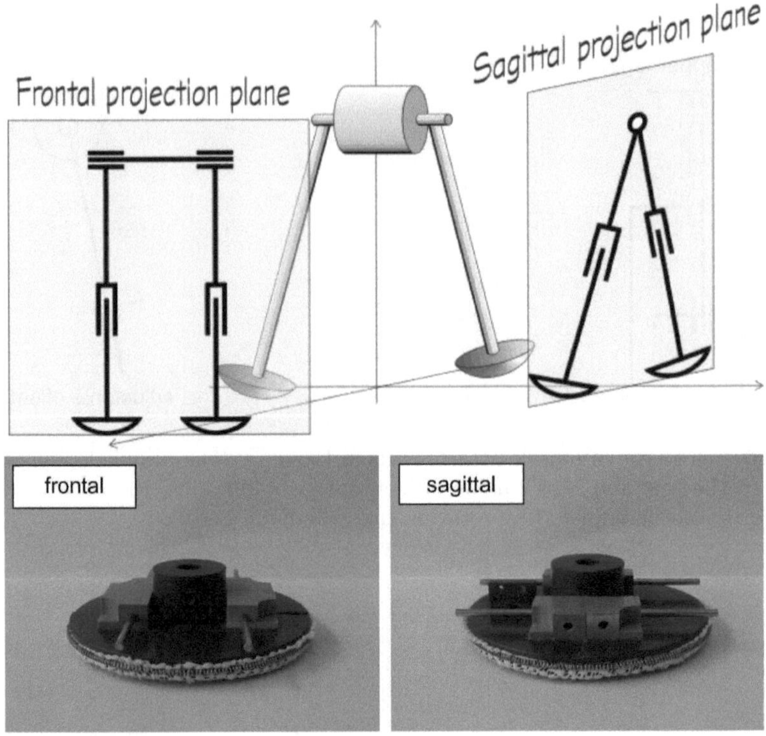

Fig. 3.12. Spatial view of the experimental biped with spherical shaped feet. The sphere radius is R=0.5 m.

3.4.4 On-Board Control

The valves are switched by reflex-like trigger signals. These reflexes are merely simple state triggers and could therefore be realized by simple analog electronics. However, the biped serves as an experimental platform so different control strategies should be easily tested. A flexible and user friendly control system is therefore desirable. Since a small control system was not commercially available, a small PC based control system was developed. SC-solutions, a specialized company in embedded control systems, took care of the complete trajectory of this development. The outcome of this development can be briefly described as follows.

A PC104 computer standard was chosen to serve as the hardware basis. PC104 has a standardized format of 90x96 [mm], and allows modular extension. The control system was composed as follows: a 486-based motherboard, a 12-bits signal processing board, a 48Mb chipdisk, and a memory extension of 32Mb. Real Time Linux (a stripped version of Slackware 4.0) served as an operating system. Software was developed that compiled Simulink®-generated files into Linux executables. With a network connection the generated Linux executables

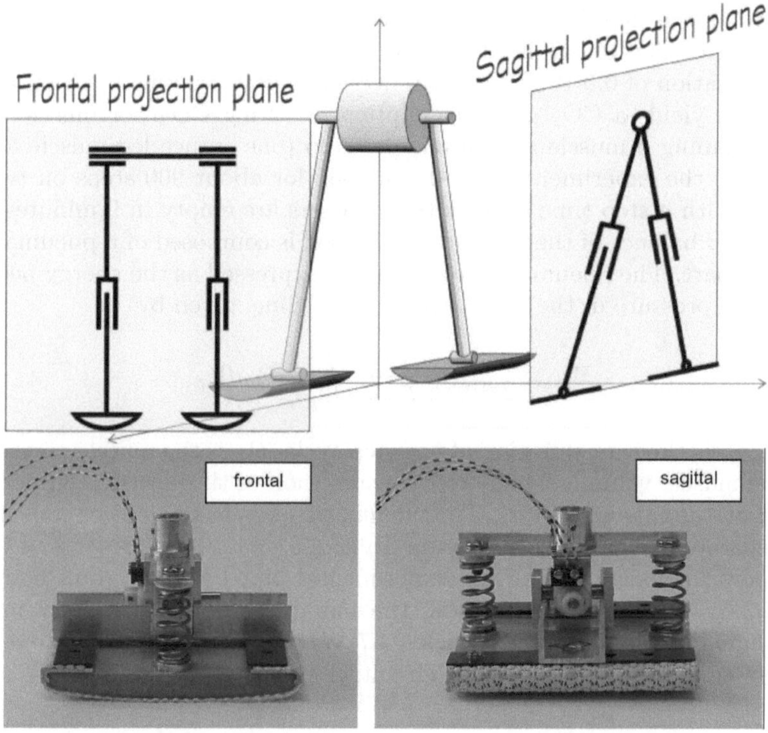

Fig. 3.13. Cylindrical shaped feet. To keep balance in the frontal plane it rocks from side to side. The cylinder radius is R=0.5 [m]. For sagittal motions, the ankle joint is free to rotate. The result of the ankle springs is a forward travel of the center of pressure during a step, similar to the forward travel of the contact point in the case of spherical feet.

are transferred from an arbitrary network PC onto the Linux control system. Real-time execution and data logging of final versions of control structures were realized with a sample frequency up to 100Hz. A stable power supply with two Li-Ion batteries was built in order to guarantee autonomous operation. Power MOSFETs amplify the valve signals generated by the computer.

3.4.5 Power Consumption

The electric power supply consists of two Li-Ion batteries of 7.2 V, 1600 mAh. Based on the electric power consumption of 14.7 W nominal the batteries can operate for 1.5 hours. After that period of time they need to be recharged. The pneumatic power supply consists of two CO_2 canisters of 16 gram each. An estimate of the amount of gas mass used per cycle Δm can be obtained with the ideal gas law:

$$\Delta m = \frac{\Delta P V_m}{R_{CO_2} T} \tag{3.16}$$

where ΔP is the muscle pressure rise, V_m is the muscle volume, R_{CO_2} is the gas constant of CO_2 and T is the temperature. Assuming a pressure rise during muscle activation of $0.5{\cdot}10^5$ Pa and an average muscle volume of 13 ml at room temperature yields a CO_2 gas consumption of 12 mg CO_2 per muscle per cycle. By assuming 3 muscle activations per step (one stance leg muscle and two hip muscles) the experimental biped can walk for about 900 steps on two CO_2 canisters. With a step time of 0.6 s the canisters are empty in 9 minutes.

The power balance of the experimental biped is composed of a pneumatic and an electric part. The pneumatic power can be expressed as the energy needed to increase the pressure in the muscle per unit of time, given by

$$W_{pneumatic} = \frac{1}{T_{cycle}} \oint_{cycle} P_m dV_m \qquad (3.17)$$

where T_{cycle} is the time duration of one step cycle, P_m is the muscle pressure and V_m is the muscle volume. The power consumption of the electric components is specified by the manufacturers. The pneumatic and electric power consumption of the separate components is given in Table 3.6.

The power consumption of the experimental biped per kilogram body mass yields 4.7 W/kg. For a comparison, the metabolic power per body mass for normal human walking [83] also yields 4.7 W/kg. According to Table 4.3, 94% of the power is used by the electric circuitry!

Table 3.6. Power consumption of the components of the autonomous biped Baps

medium	component	peak power [W]	average power [W]
electric	PC	17.50	12.00
	signal board	1.60	1.60
	1 gyroscope	0.02	0.02 (3x)
	1 valve	0.75	0.28 (4x)
	total		**14.78**
pneumatic	hip muscles		0.5
	leg muscles		0.4
	total		**0.9**

3.4.6 Walking Results

Baps has shown successful autonomous operation. This confirms the proper operation of all the mechanical, pneumatic, electric, and software components as parts of a fully self-supported system. To test the operation, Baps has walked series of up to 10 steps as shown in Fig. 3.14. These experiments were done with a simple controller that triggered the hip muscles once per step based on the signal of the frontal gyroscope. These results conclusively demonstrated that it is possible to create autonomous walking machines based on the concept of passive dynamic walking and powered with pneumatic McKibben muscles. Thus, we decided to continue this line of research.

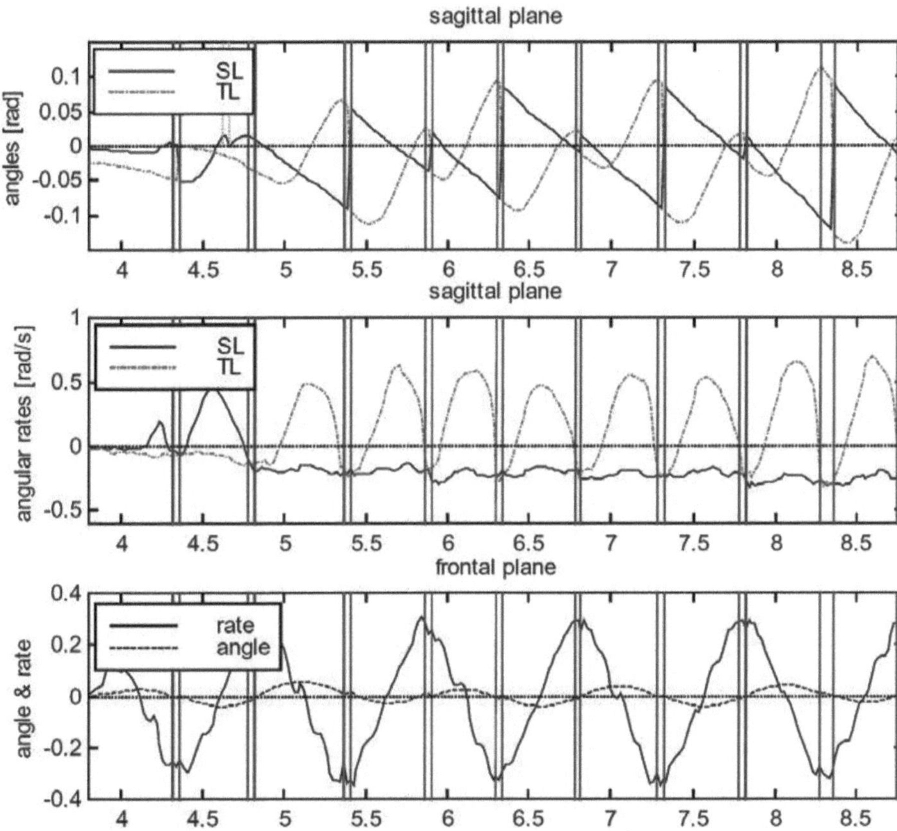

Fig. 3.14. The measured states of Baps during autonomous walking experiments with a simple controller. The direction of rotation is defined to be positive when the foot is in a forward position. The vertical lines approximately indicate the double stance phase. The integrator of the frontal gyroscope is reset every step, so that the frontal angle (third graph) is reasonably reliable. The sagittal angles are not reset each step, resulting in large drift (top graph). We show these integrated angle estimates for easy interpretation. The sample rate during the experiment was 50 [Hz].

However, we did not manage to obtain stable, sustained locomotion. The side-to-side rocking motion and the forward stepping motion were not sufficiently synchronized, and after approximately 10 steps the phase shift became too large. We have attempted to implement a more advanced controller that used a two-dimensional model of the legs to predict the instant of heel strike. The predictions and measurements are shown in Fig. 3.15. The figure shows that the predictions differ much from the actual behavior. The problem is that the model is two-dimensional, while the actual behavior is three-dimensional. Motions in the other planes (frontal plane and top view plane) significantly influence those

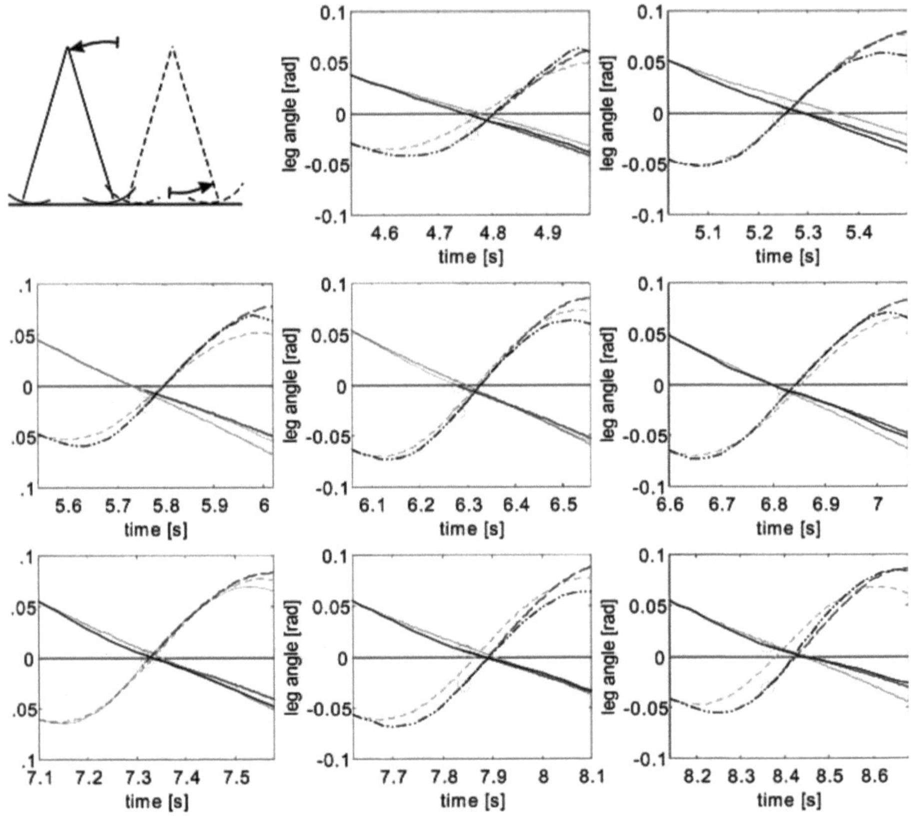

Fig. 3.15. Leg angles of eight steps. Prototype measurements are black lines. The model simulations are performed with two sets of initial conditions, one at toe-off (grey lines), and one at mid-stance (light grey lines and circles). A solid line depicts the stance leg angle, and a dash-dotted line depicts the swing leg angle. Clearly, the model prediction at mid-stance is more accurate than the model prediction at toe-off, but both show significant differences with the actual motion.

in the sagittal plane. We realized that at that time we did not have a sufficient understanding of the 2D sagittal motions, let alone the full 3D walking motions. Therefore, the next robot in this book is deliberately made simpler by giving it two-dimensional dynamic behavior.

4. Mike; How to Keep from Falling Forward

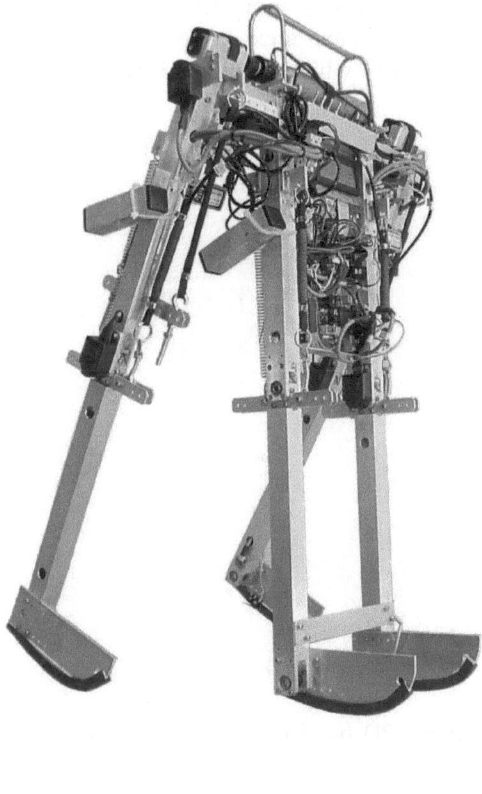

M. Wisse and R.Q. van der Linde: Delft Pneumatic Bipeds, STAR 34, pp. 45–62, 2007.
springerlink.com

4.1 Introduction

Purely passive walkers (Chapter 2) and our first prototype Baps (Chapter 3) can walk efficiently, but they fall too easily. Now it is time to do something about it. The tools are here: we now have knowledge of passive dynamic walking and we have a way of actuating these walkers with pneumatic McKibben muscles. How can the actuation be used to enhance the stability of the walking motion? This chapter will provide an answer, first by analyzing the stability of an unactuated model, followed by an implementation of hip actuation both in an elementary simulation model and in the prototype 'Mike' (Fig. 4.1) that was built for this purpose. The content of this chapter is based on two of our papers [141, 184],

It is too complex to solve the full 3D stability problem at once, so in this chapter we will focus on 2D models and prototypes that cannot fall sideways. In simulations, this means that we simply don't model the third dimension. However, in prototype studies it is more difficult to ignore the existence of a third dimension. Following McGeer's example [108], the prototype is built with a double pair of legs (Fig. 4.1) which should minimize sideways dynamics.

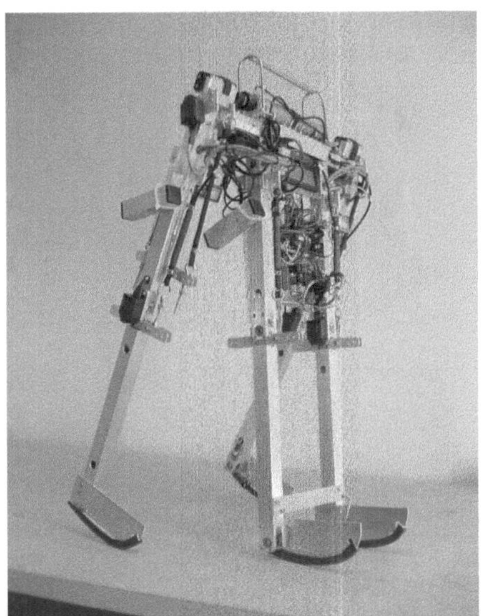

Fig. 4.1. Prototype Mike; a 2D passive dynamic walking robot with pneumatic McKibben muscles at the hip

4.2 Modeling and Analysis

4.2.1 The Simplest Walking Model

This research starts with the simplest walking model [54] as described in Chapter 2. We described how a cyclic walking motion can be found and how it can be analyzed using the eigenvalues of the linearized 'Stride Function', Eq. (2.26). This analysis tells us whether a cyclic motion is stable or not, but it is only valid for very small disturbances because of the linearization.

4.2.2 Basin of Attraction

For practical use, one wants to know when the walker keeps walking, and when it falls down. Clearly, the wider the range of initial conditions of $[\theta, \dot{\theta}]$ that result in continuous walking, the more tolerant is the walker for incorrect launches and in-motion disturbances. The entire collection of initial conditions leading to walking is what is called the *basin of attraction.* We know that there must be some basin of attraction when the walker is linearly stable around the fixed point, but how large is it? Here we describe how to find the complete basin of attraction with the cell mapping method [78].

The region of feasible initial conditions is subdivided into a large number (N) of small cells. All unfeasible initial conditions (such as a backward motion, i.e. $\dot{\theta} > 0$) are regarded as a small number (z) of very large cells, so called *sink cells.* The cells are numbered 1 to N+z. By application of the stride function **S** to the center of each cell, all of the N+z cells point to initial conditions inside one of the other cells, except the sink cells which point to themselves by definition. Starting with cell 1, a sequence of cells appears by following the pointers. This sequence either ends in a sink cell or in a repetitive cycle. This cycle can consist of one self-repeating cell (a fixed point), or a number of cells (multiple-period walking, Garcia *et al.* [53]). The fixed point is identified and all cells in the sequence are labeled as the basin of attraction of that fixed point. Then the procedure is repeated with cell 2, then cell 3, etcetera. As soon as a known cell (from a previous sequence) is encountered, the current sequence merges with that of the known cell. The procedure is repeated until all cells are labeled.

Note that the calculation of the basin of attraction is a costly business as the number of calculations increases with the power of the number of degrees of freedom. This is one of the reasons that we perform the simulation analysis on the simplest walking model instead of on the more complete model of Mike.

4.2.3 Failure Modes

Application of the cell mapping method results in a list with all attractors (fixed points) and classification of all discretization points into this list. Not only period-one walking gaits can be found, also period-k walking gaits. Results of the cell mapping method are as accurate as the discretization, within these tolerances fixed points may come and go. For example, what appears to be a fixed cell might in fact be slowly changing initial conditions (smaller changes than the discretization) of subsequent steps.

With the cell mapping method, the basin of attraction of the simplest walker (at a slope of $\gamma = 0.004$ rad) is calculated in [141], see Fig. 4.2. The basin of Attraction is represented by the very thin area A, otherwise the walker falls

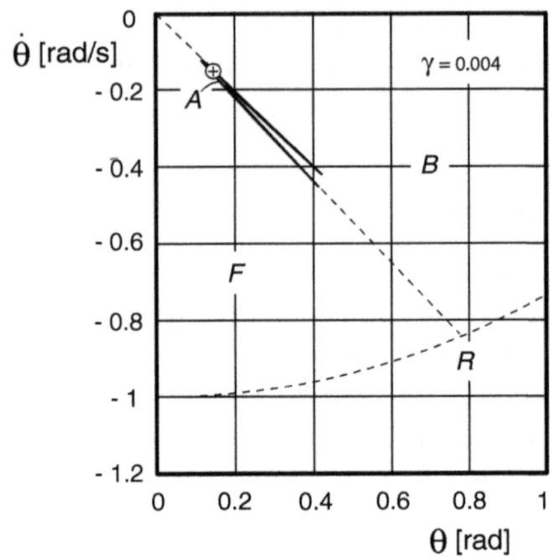

Fig. 4.2. Poincaré section for the simplest walker with initial stance leg angle θ and velocity $\dot{\theta}$ together with failure modes; falling Forwards, falling Backwards and Running, and the basin of Attraction of the cyclic walking motion $(\theta, \dot{\theta}) = (0.1534, -0.1561)$ (indicated with '+') at a slope of $\gamma = 0.004$ rad.

Forwards or Backwards, or, if started with high speeds, the stance foot loses compressive ground contact (Running).

4.2.4 Behavior Inside the Basin of Attraction

From Figure 4.2 it is obvious that the basin of attraction is only a very small region. For better insight in the shape, Figure 4.2 is enlarged and sheared, leading to Figure 4.3. The vertical axis now represents the sum of the stance leg angle and scaled angular velocity. The horizontal line at $\theta + \dot{\theta} = 0$ corresponds with the '-45 degree'-line in Figure 4.2. Figure 4.3 is obtained with application of the cell mapping method with a discretization of about 200×250 points ($\Delta\theta = 0.002$ rad, $\Delta(\theta + \dot{\theta}) = 0.0002$ rad), the solid lines are a manual continuous interpretation of the discrete simulation results for the boundary of the basin of attraction.

Figure 4.3 shows that the small basin of attraction is mostly bounded by falling backward on one side and falling forward on the other side. The basin of attraction seems to be a continuous and tailing area. The different areas show fractal-like entanglement. We will discuss the behavior of the walker in these areas by going over a vertical line, $\theta = 0.2$ rad, in Figure 4.3 from area F to area B, crossing the basin of attraction at least four times. Point ($\theta = 0.2$ rad, $\dot{\theta} = -0.23$ rad) lies in area F. If started with such initial conditions, the walking model will fall forward; the swing leg is allowed to pass through the floor to ignore the otherwise inevitable foot-scuffing, but does not rise above floor level anymore. Going up, the area changes from falling Forward to the basin of Attraction. Just before crossing this boundary, the behavior changes. Not the first step after these initial conditions is failing, but the model first walks some steps before eventually falling forward. The closer to the basin of attraction,

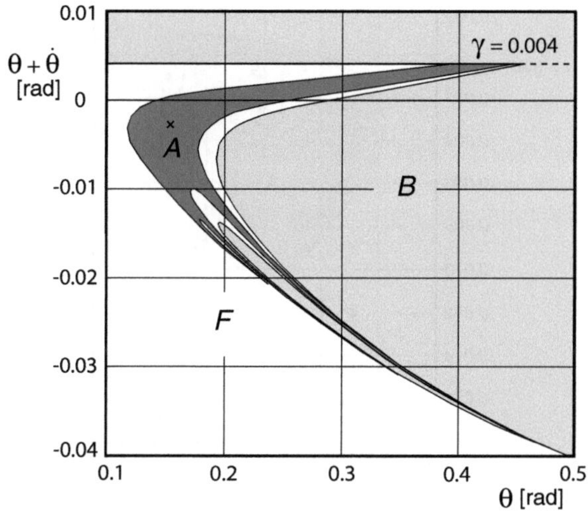

Fig. 4.3. Poincaré section for the simplest walker, enlarged and sheared section from Figure 4.2, together with failure modes; falling Forward, falling Backward, and the basin of Attraction of the cyclic walking motion.

Table 4.1. Initial conditions of a number of subsequent steps, started just inside the basin of attraction and going to the fixed point

step	θ rad	$\dot{\theta}$ rad	$\theta + \dot{\theta}$ rad
1	0.2000	-0.2165	-0.01645
2	0.1788	-0.1917	-0.01290
3	0.1756	-0.1841	-0.00850
4	0.1878	-0.1888	-0.00100
5	0.1586	-0.1599	-0.00134
6	0.1459	-0.1488	-0.00295
7	0.1492	-0.1526	-0.00337
8	0.1539	-0.1569	-0.00302
9	0.1558	-0.1583	-0.00256
⋮	⋮	⋮	⋮
f.p.	0.1534	-0.1561	-0.00269

the more steps it takes before failure. If started in the first tail of the basin of attraction, encountered when going up, the walker will eventually settle into steady cyclic walking with initial conditions of the fixed point. The path toward the fixed point is presented in Figure 4.4 and in Table 4.1. The motion of the legs is shown in Figure 4.5. After nine steps, the walker is close to the fixed point, and continuing the simulation will show an asymptotic approach.

Even more up on the line ($\theta = 0.2$), area F is encountered again. Starting there leads to falling forward after a number of steps. In this manner, A, B,

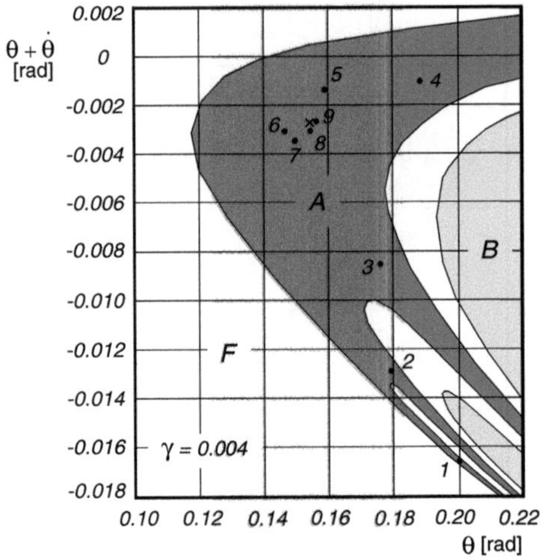

Fig. 4.4. A number of steps, started just inside the basin of attraction and going to the fixed point. Together with failure modes; falling Forward, falling Backward, and the basin of Attraction of the cyclic walking motion.

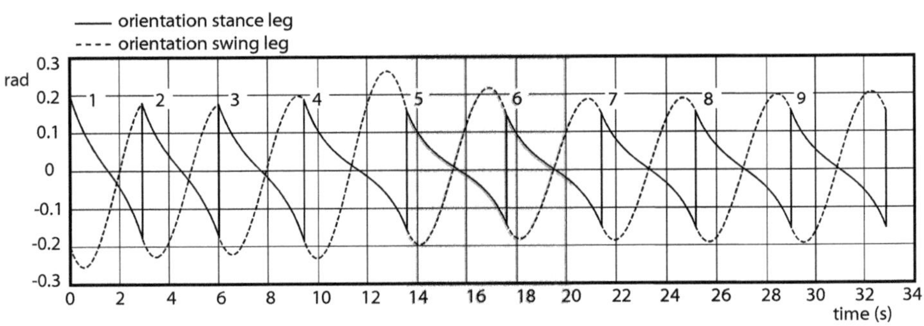

Fig. 4.5. Orientation of the stance leg, solid line, and swing leg, dotted line, with respect to the normal on the slope for a walker started just inside the basin of attraction and going to the fixed point

and F are crossed several times, until we reach $\theta + \dot{\theta} = \gamma$. Above this boundary, the stance leg will not reach midstance and fall backward. In general, if started inside the basin of attraction, the initial conditions spiral toward the fixed point. If started just outside the basin of attraction, the walker will take a few steps but eventually fail. The further away, the smaller the amount of successful steps before failure.

4.2.5 Basin of Attraction Versus Slope Angle

The size of the basin of attraction determines the amount of disturbance that the walker can handle without falling. The stability of the fixed point determines if, and how fast the walker recovers from a small disturbance. The latter analysis is less time-consuming and therefore very useful to determine the existence of a basin of attraction. The applicability of the walker however depends on the allowable size of disturbances. Therefore, we investigate the dependency of the basin of attraction on the only model parameter, the ramp slope γ. Figure 4.6 shows the development of the basin of attraction for an increasing ramp slope γ. As the slope increases, the basin of attraction decreases in size and gets more and more tails at the boundaries, which appear to be fractal-like. At and beyond a slope angle of 0.019 rad the basin of attraction has vanished, leaving fractal-like boundaries between the regions of falling forward and falling backward.

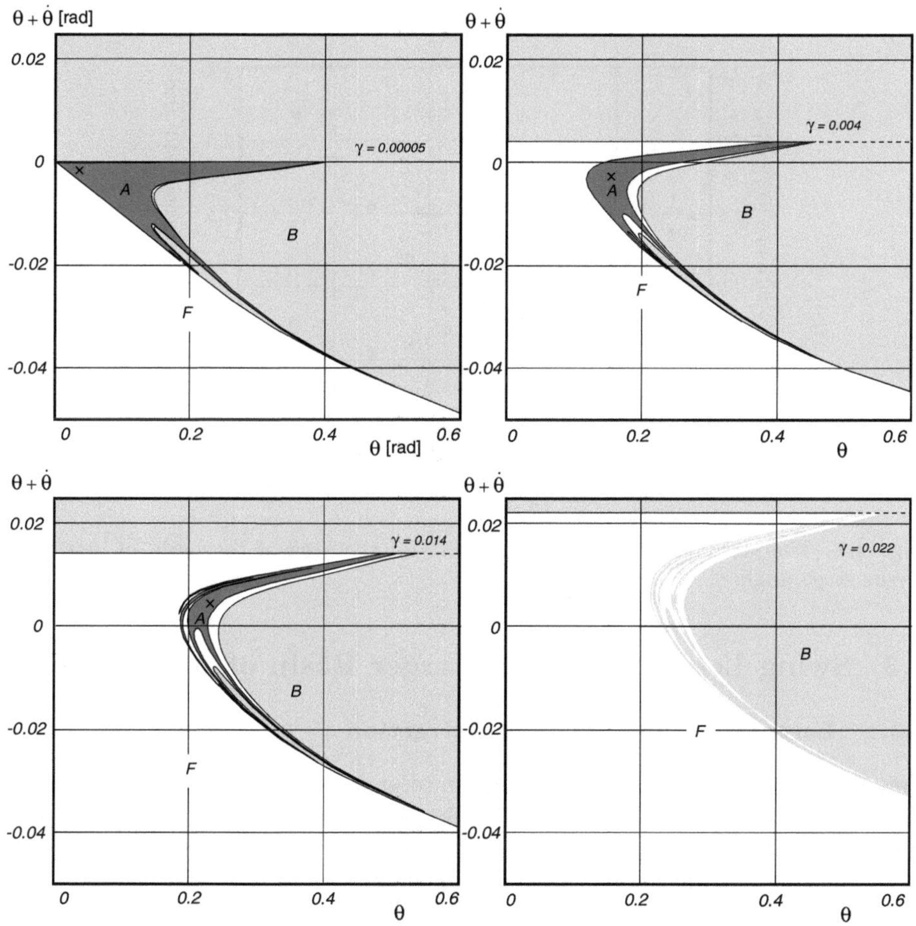

Fig. 4.6. Development of the basin of attraction with increasing floor slopes, together with failure modes; falling Forward, falling Backward, and the basin of Attraction of the cyclic walking motion

4.2.6 Basin of Attraction Versus Eigenvalues

We compare the size of the basin of attraction with the linearized stability of the cyclic motion, see Figure 4.7. The stability is measured as the largest of the two eigenvalues (modulus) of the linearized stride function (2.26). As stated by Garcia *et al.* [53], for $0 < \gamma < 0.0151$ rad the period-one gait is stable. For higher slopes, only higher-period gaits are stable, having a small basin of attraction. The eigenvalues would lead to believe that a slope $\gamma = 0.012$ rad would be preferable. However, the basin of attraction, measured as the number of cells inside the basin of attraction times the area of one cell, is not at its maximum. Clearly, there is no direct relation between the stability of the cyclic motion and its basin of attraction.

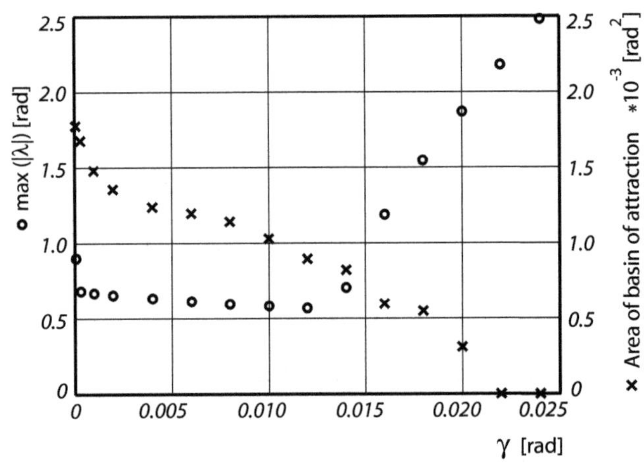

Fig. 4.7. The stability of the cyclic motion and the area of the basin of attraction versus slope angle γ

4.3 Swing Leg Action for a Larger Basin of Attraction

4.3.1 Largest Possible Basin of Attraction

Now that the basin of attraction of the simplest walker is known, some questions arise. Is it a sufficiently large basin of attraction, or is control necessary? Can control of only the swing leg have any substantial positive effect on the basin of attraction? Before answering these questions, we should recognize that any conceivable hip action does not influence the current stance leg motion whatsoever, because of the fact that the swing leg has an infinitesimally small mass. The only thing that matters is the hip angle at heel strike. In other words, swing leg control of the simplest walker can only influence the *subsequent* step.

Therefore, swing leg control cannot do anything for the walker if the current hip velocity is not enough to pass the apex of the hip trajectory. The mathematical equivalent of this requirement is the following energy inequality:

$$\frac{1}{2}M(\dot{\theta}l)^2 > Mgl(1-\cos\theta) \tag{4.1}$$

or, rewriting and scaling M, g, and l to unity:

$$|\dot{\theta}| > 2\sin\frac{\theta}{2} \tag{4.2}$$

This inequality is represented in Fig. 4.8 with Line (1). Note that this is the familiar separatrix in the normal simple pendulum phase portrait. Also note that Line (1) does not coincide with the dashed boundary between area B and area F in Fig. 4.2. The area between the two lines represents a set of initial conditions that, for the fully passive walker, does not lead to *immediate* falling backward, but to a short series of successful steps that *eventually* leads to falling backward [141].

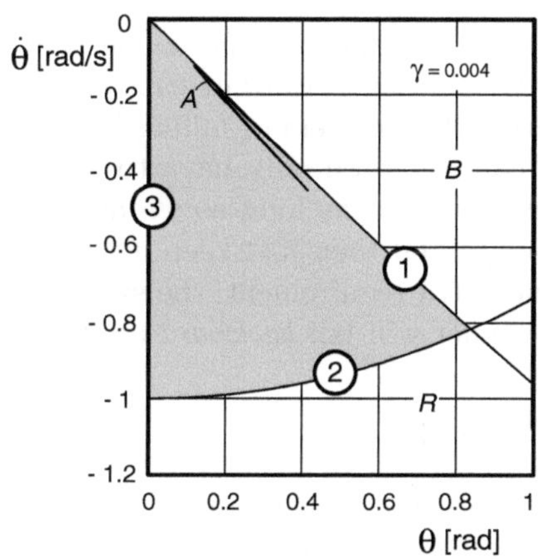

Fig. 4.8. Maximally obtainable basin of attraction (gray area, bounded by Lines 1, 2 and 3, see text) and uncontrolled basin of attraction (thin area A) of the simplest walking model. The entire problem of falling forward can be solved with swing leg control, while the problem of falling backward (area B) remains existent and would need something else than swing leg control.

The second boundary to the initial conditions for making a successful step is the requirement for compressive foot contact: at high velocities the centrifugal effect overcomes gravity and the stance foot would loose contact. This boundary is represented in Fig. 4.8 as Line (2). The scaled vertical contact force f_v is given in Eq. (2.9) as:

$$f_v = -\cos(\theta)(\dot{\theta}^2 - \cos(\theta-\gamma)) \tag{4.3}$$

Requiring that $f_v > 0$, we get the inequality for Line (2):

$$|\dot{\theta}| < \sqrt{\cos\theta} \tag{4.4}$$

Note that the fact that Line (2) passes through $\{\theta, \dot{\theta}\} = \{0, -1\}$ corresponds to saying that for this nondimensionalized walker the Froude number $\sqrt{v^2/(gl)}$ is equal to 1 (with body speed v equal to stance leg velocity $\dot{\theta}$ due to the unit leg length).

The third and last boundary of the maximally achievable basin of attraction is somewhat arbitrary; we only investigate initial conditions with a positive stance leg angle, i.e. starting with the stance foot in front of the swing foot, resulting in Line (3).

Lines (1), (2), and (3) are the outer boundaries for any basin of attraction that swing leg control could possibly achieve for the simplest walker. Therefore we choose to use the area inside these boundaries as a reference for the size of the basin of attraction. Comparing the area of the thin region A in Fig. 4.2 with the large gray area in Fig. 4.8, we find that without control, the basin of attraction is only 0.3% of the maximally achievable area. This result justifies the search for a controller to enlarge the basin of attraction.

Summarized so far, we have said that the uncontrolled simplest walker very rarely walks. It mostly falls, either forward or backward. When adding swing leg control, we cannot address the problem of falling backward whatsoever. Therefore, for the swing leg controller we only have to consider the forward falling problem. This makes things easy; as long as we make sure that the swing leg swings forward fast enough, and then just keep it there, the problem should be solved. There is only one extra requirement; the swing leg should not be too far forward, otherwise the walker will fall *backward* at the subsequent step.

4.3.2 The Rimless Wheel

If for now we stick to the idealized situation of a massless swing leg, we could imagine the swing leg controller putting the swing leg at a preset, constant, forward angle immediately after the step has started. The behavior of the walker would then be exactly equal to the 'rimless wheel' (Fig. 2.2), a system that has a number of 'legs' (the spokes) at equal, fixed angles (see the inset in Fig. 4.9). This system has been studied in depth by Coleman [31], who concluded that it would always reach a stable cyclic walking motion, provided that the leg angle is small enough for the floor slope angle. For the slope angle of 0.004 rad that we use throughout this chapter, the inter-spoke angle should be smaller than 0.4 rad [30], otherwise it would slow down and eventually come to a stop. We choose a safe value of 0.3 rad, i.e. a stance leg angle $\theta = 0.15$ rad at the start of a step, which corresponds to the natural gait of the simplest walking model at this slope (see Fig. 2.7). Fig. 4.9 shows the basin of attraction of the rimless wheel with this inter-spoke angle of $2 \cdot 0.15$ rad.

The fixed inter-spoke angle makes it impossible to start the rimless wheel with a larger initial leg angle than 0.15 rad, unless it were started at the edge of a table top. But even then, it would converge to its fixed point. Other than that, the only important gap in the basin of attraction is the small corner in the top, bounded by a line of constant energy through the point where Line (1) crosses $\theta = -0.15$ rad. In that corner, the initial energy of the rimless wheel is enough to make it through the first step, but the fixed inter-spoke angle is too large to make it through the second.

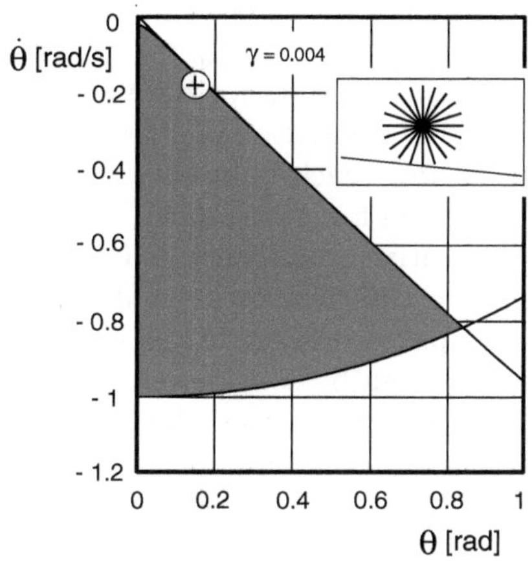

Fig. 4.9. Basin of attraction of the rimless wheel (model: see inset) with an inter-spoke angle of 0.3 rad. The '+' indicates the fixed point of this model at a slope of $\gamma = 0.004$ rad.

In conclusion, only considering swing leg control, the maximally obtainable basin of attraction can be achieved with a controller mimicking the rimless wheel.

4.3.3 A Realistic Actuation Model

The previous section indicates that the stability behavior of the rimless wheel is very close to the maximum achievable with swing leg control. So, now it is time to devise a simple but realistic form of actuation that mimics the rimless wheel behavior, acknowledging that instantaneous leg positioning is impossible. We propose to use a spring and a damper at the hip joint with a variable setpoint which can provide for an extra internal torque T to the swing leg. The equations of motion for this model are the same as those of the simplest walking model (Eq. 2.8), extended with torque T resulting in:

$$\begin{bmatrix} \ddot{\theta} \\ \ddot{\phi} \end{bmatrix} = \begin{bmatrix} \sin(\theta - \gamma) \\ \sin(\phi)(\dot{\theta}^2 - \cos(\theta - \gamma)) + \sin(\theta - \gamma) + T \end{bmatrix} \tag{4.5}$$

with

$$T = -k(\phi - \phi_{sp}) - c\dot{\phi} \tag{4.6}$$

The setpoint ϕ_{sp} is set to 0.3 rad corresponding to the fixed point of the passive walker at a slope of $\gamma = 0.004$ rad. The stiffness k is the parameter that we vary, where $k = 0$ corresponds to the fully passive simplest walker and $k = \infty$ corresponds to the rimless wheel. Note that k is scaled with the foot mass m. The damping factor c is set as a function of k to provide critical damping:

$$c = 2\sqrt{k} \tag{4.7}$$

Note that a physical realization of this type of actuation requires an active shift of the setpoint after each heel strike from ϕ_{sp} to $-\phi_{sp}$ or vice versa.

Fig. 4.10 presents the stability results for different stiffness values. A higher stiffness results in a faster swing leg motion and thus provides a better resistance against falling forward. The drawback is in energy consumption, but unfortunately the model with its massless feet is too much a simplification of real walking machines to allow for quantitative statements on energy expenditure. With this active hip spring stiffness we can arbitrarily make the basin of attraction as large as necessary up to complete coverage of the maximally obtainable area, so the problem of falling forward can be considered to be solved. Moreover, this is achieved without any feedback control other than a setpoint shift at heel strike.

Remember that this is just one possible way of speeding up the swing leg. It is not this particular implementation that counts, but the main idea behind it that the swing leg should be swung forward quickly, and then kept there at a not-too-large leg angle. We should emphasize here that, although simulated with a floor slope $\gamma = 0.004$ rad, this control rule works equally well for any slope larger than

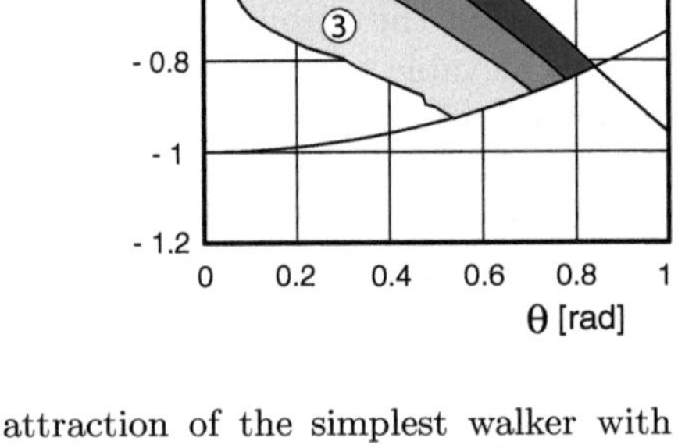

Fig. 4.10. Basin of attraction of the simplest walker with active hip spring. The setpoint of the hip spring is $\phi_{sp} = 0.3$ and critical damping is applied. The higher the hip spring stiffness, the larger the basin of attraction; $k = 25$ leads to area (1), $k = 50$ leads to areas (1) plus (2), and $k = 100$ leads to areas (1) plus (2) plus (3). The fixed point is for all three stiffness settings approximately the same, located at the '+'. A disturbance from a step down in the floor would result in initial conditions away from the fixed point in the approximate direction of the white arrow.

that (for smaller slopes, the preset leg angle should be decreased accordingly). Moreover, it appears that the amount of success of other dynamically walking bipeds can be attributed to how well their swing leg is brought forward, although the control rules are usually formulated implicitly (e.g. Westervelt *et al.* [171]).

4.4 Prototype Design and Experiments

4.4.1 Prototype Construction

We applied the proposed swing leg control to our prototype 'Mike' (Fig. 4.1). Mike weighs 7 kg and measures h x b = 0.7 x 0.4 m, and it uses 0.4 grams of compressed CO_2 per second. Movie clips of Mike in action are available at our web site [1]. Mike has four legs symmetrically paired, giving it approximate 2D behavior. It differs from the simplest walking model by having knees, a distributed leg mass, round feet and by walking on a level floor (no slope!).

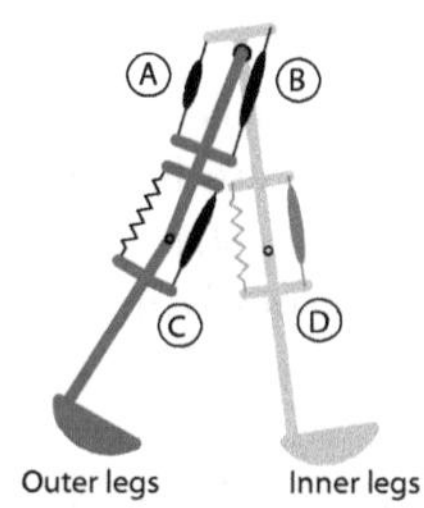

Fig. 4.11. Schematic structure and muscle attachments of Mike

Mike is actuated with a total of eight McKibben muscles. The McKibben muscles are arranged according to Fig. 4.11. The hip joint is actuated with two muscles (A) and two muscles (B) which are arranged as two antagonistic pairs, providing a combined joint stiffness. The knees are actively extended with McKibben muscles (C) and (D) which are counteracted by weak passive springs. There is no ankle actuation; the arc feet are rigidly attached to the shanks.

4.4.2 Actuation System

The McKibben muscles are fuelled from a 5.8 MPa CO_2 container via a two-stage pressure regulator and via electromagnetic valves that are activated by switches underneath the feet. The second-stage pressure regulator output is manually adjustable between 0.1 and 0.6 MPa resulting in a hip joint stiffness up to 5 Nm/rad and a damping somewhat less than critical damping (estimated by observation). It is not feasible to perform a proper mapping between this stiffness in Mike and the scaled stiffness in the simplest walking model due to the extensive differences between the two, such as leg mass, foot arc radius, muscle non-linearities and significant air flow dynamics. Therefore the comparison between the two will be of a qualitative nature only.

If a valve is switched 'on', the muscle is filled from the pressure regulator output; if switched 'off' it empties into atmosphere. For example, at the instant of contact of the inner leg foot switch, the outer knee muscles (muscle C in Fig. 4.11) are switched 'off' to allow this knee to bend. A manually tuned 400 ms later they are switched back 'on', ensuring a properly extended knee for the next step.

The proposed swing leg control in Eq. 4.6 is implemented by alternating the states of the antagonistic hip muscles. When the foot switch of the inner legs is activated, muscle B in Fig. 4.11 is switched 'on' and muscle A is switched 'off'. At the next step this is inverted. As a result, the hip joint has a constant stiffness but a setpoint that alternates between ϕ_{sp} and $-\phi_{sp}$. The joint stiffness can be adjusted without altering the setpoint. We want to emphasize that there is no feedback control other than this once-per-step switching between preset muscle pressures.

4.4.3 Stability Results

Mike walks well. Fig. 4.12 shows a sample of the sustained gait for a hip muscle pressure of 0.55 MPa, see [1] for video evidence. We would have liked to create a figure of its basin of attraction like Fig. 4.10. However, the combined limitations on the number of experiments to be performed and on the physical possibilities to create controlled disturbances have led us to concentrate on one representative disturbance, namely a step-down. In the experiments the prototype walks steadily and then takes a step down of increasing height, see Fig. 4.13. Such a step down results in a larger stance leg velocity at the subsequent step as sketched with the white arrow in Fig. 4.10. The larger the step down height, the larger the arrow. If a larger hip muscle stiffness indeed allows a bigger step down, then our swing leg control rule is validated.

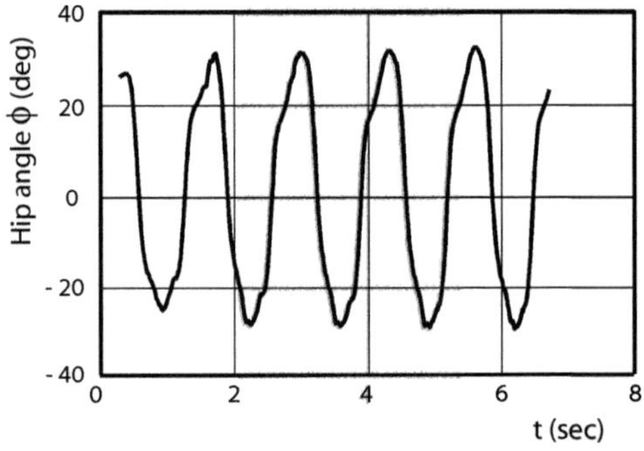

Fig. 4.12. Typical walking result with active hip muscles (0.55 MPa) on a level floor. The prototype completes 10 steps in this trial, showing convergence towards its fixed point after a manual launch.

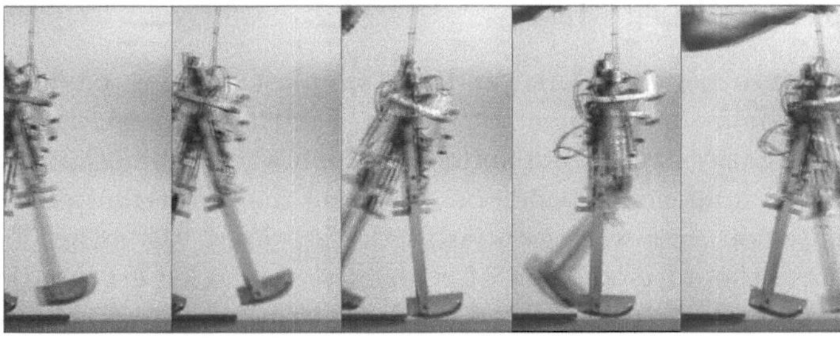

Fig. 4.13. Experiment with Mike walking on level floor and taking a step down as a representative disturbance

The stability results are shown in Fig. 4.14. A hip muscle pressure lower than 0.35 MPa did not provide stable walking at all, not even without disturbances. When the pressure was increased, a larger step down could be handled. The muscles cannot sustain pressures higher than 0.55 MPa. Fig. 4.14 clearly shows a better robustness against falling forward with a higher hip pressure which corresponds to a higher joint stiffness.

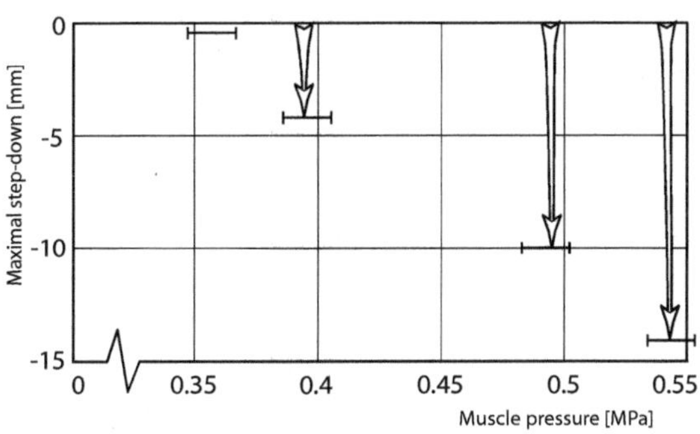

Fig. 4.14. A higher hip muscle pressure setting (corresponding to a higher hip joint stiffness) results in a larger step-down size and thus in a better resistance against disturbances. The arrows are depicted to indicate correspondence with the white arrow in Fig. 4.10.

4.5 Discussion

4.5.1 Level Floor

One of the differences between Mike and the simplest walking model is that Mike's legs are not massless. When quickly moving the swing leg forward, it is not only rotated but its center of mass is lifted a little bit. For walkers with non-massless legs, this provides a way of putting energy into the system. For Mike this is sufficient to replace the gravitational energy that the simplest walking model obtained from walking downhill. As a result, Mike can walk on level terrain.

4.5.2 Distributed Leg Mass

Another difference between Mike and the simplest walking model is that the swing leg has a non-zero moment of inertia. For the simplest walker, any conceivable hip actuation would not influence the stance leg motion. For Mike, it does. The actual influence depends on the exact mass distribution of the swing leg. If its center of mass were located at the hip joint, the swing leg acceleration would have the adverse effect of a *forward* stance leg acceleration. With a center of mass more in the middle of the leg, the swing leg acceleration induces a *backward* acceleration (i.e. a deceleration) of the stance leg, which buys the prototype some extra time to place its swing leg. A linearized dynamic analysis of a straight-legged mechanism at mid-stance shows that such advantageous deceleration occurs if

$$I < mc(l - c) \tag{4.8}$$

with the leg moment of inertia I, leg mass m, leg length l and the distance between the hip joint and the leg's center of mass c. From Eq. (4.8) we deduce that most normal constructions ($I \approx \frac{1}{12}ml^2$) result in a slight (advantageous) deceleration of the hip.

Even though the stance leg and hip are decelerated, the center of mass of the entire system is accelerated forward due to the swing leg acceleration. This provides a neat opportunity to make the prototype start from a standstill, which was implemented successfully in the prototype.

4.5.3 Feet

The simplest walking model has point feet, whereas Mike's feet have a circular shape. Such arc feet provide extra robustness against the complementary problem of falling *backward*. As depicted in Figs 4.2 and 4.8, the walker will fall backward if it has not enough velocity to overcome the vertical position. Circular feet smoothen the hip trajectory and thus relax the initial velocity requirement. As a result, the basin of attraction is enlarged in the upper right direction. We believe that arc feet are a stability improvement measure complementary to the swing leg control rule proposed in this chapter. However, a decisive study on the effect of arc feet on the basin of attraction has yet to be performed.

4.5.4 Knees and Muscles

Other differences between Mike and the simplest walking model are the knees and the nonlinear muscle properties. The knees do not essentially change the global behavior [109] as the swing leg starts and ends fully extended. The muscles do not behave in an exactly linear fashion. Especially the damping and friction losses are strongly dependent on the muscle length, providing much more resistance when the muscle is close to its maximal extension. For Mike, this behavior is useful. The swing leg is brought forward without much resistance and is then effectively slowed down by the elongating hip muscle.

4.5.5 Human Walking

From measurements on human walking (e.g. Inman *et al.* [83], Winter [173]), it has been found that there is hip muscle activity that accelerates the swing leg at the beginning of the swing phase and decelerates it towards the end of the swing phase. Selles *et al.* [142] confirms that the human swing leg moves faster than it would if it were purely passive. At first sight, this seems like a waste of energy, but Kuo [97] hypothesizes that humans might speed up their swing leg to actually *improve* walking energetics, as it indirectly reduces the energy loss at heel strike. In addition to Kuo's hypothesis, the research in this chapter provides a second hypothesis to why humans might speed up their swing leg; it increases the robustness against disturbances. It is not relevant what the exact muscle patterns are: as long as they result in an acceleration of the swing leg to a forward position, the positive robustness effect is similar to that of the control algorithm presented in this chapter.

When comparing our proposed swing leg control rule to human muscle activation in more detail, it should be noted that our controller is always active, even if no disturbance is present. A more efficient controller would monitor (or even predict) the size of the disturbances and adjust the applied torques correspondingly. It seems reasonable to compare this to a walking human who will violently throw forward his swing leg as a reflexive reaction to tripping. We hypothesize that this reflex is a 'full power' version of the stabilizing control rule presented in this chapter, while in normal walking a 'low power' version is applied. This hypothesis is in line with Pratt [129] and Forner Cordero [38] who suggest that the maximum walking speed in humans is dictated by how fast one can swing his or her leg.

4.6 Conclusion

This chapter started with the question how hip actuation can be used to enhance the stability of the walking motion. The answer is two-staged.

First, we showed that swing leg control can only address the problem of falling *forward*. If the simplest walker falls *backward*, there is no way that any swing leg control can change this; there is simply not enough energy in the system to move past the vertical position.

Second, we showed that a simple controller can completely solve the problem of falling forward; all it needs to do is to get the swing leg timely in a forward position. A damped hip spring with a forward setpoint already suffices. The specific control and actuation details are not important as the same result can be achieved with any controller if it is based on the following rule: *"You will never fall forward if you put your swing leg fast enough in front of your stance leg. In order to prevent falling backward the next step, the swing leg shouldn't be too far in front."* A controller designed according to this rule is easy to implement, because no a-priori knowledge of the passive dynamic walking motion is needed.

We validated this rule with experiments with an autonomous, two-dimensional (four-legged) prototype with knees. The hip joint was actuated with McKibben muscles which provide a joint stiffness proportional to their internal CO_2 pressure. By using only one muscle of a pair of antagonistic muscles, the hip joint was given a stiffness and a forward setpoint each step. In this manner, the swing leg was accelerated forward according to our proposed control rule. The prototype was made to take a step-down during a steady walk, and the maximal step-down height was recorded as a function of the hip muscle pressure (hip joint stiffness). It was shown that a higher pressure indeed allows a higher step-down. The resultant robust gait can be viewed at [1].

5. Max; Adding the Upper Body

M. Wisse and R.Q. van der Linde: Delft Pneumatic Bipeds, STAR 34, pp. 63–83, 2007.
springerlink.com

5.1 Introduction

More than two thirds of the mass of a human being is located in the upper body (head-arms-trunk), while the legs are relatively lightweight. The addition of an upper body to passive dynamic walkers remains an active research topic; so far only the legs have been considered. The problem is that the upper body should be stabilized in the upright position, while at the same time the alternating swing leg should be able to swing passively to a forward position. Some passive solutions have been found [17, 58], in which the upper body is another passive pendulum-like component. Also a number of active control solutions have been proposed, such as McGeer's 'levered isotonic tendons' [110], variable springs [160] or a controllable 'backlash clutch' in the hip joints [122], all fairly complex solutions. In contrast, we have searched for an alternative, mechanical solution.

In this chapter, we propose to design the hip joint as a passive bisecting mechanism, similar to that in a pair of compasses. We will first analyze a simple point mass model in Sections 5.2, 5.3 and 5.4. Next, Section 5.5 will present the two-dimensional simulation model and prototype (Fig. 5.1) developed for this study. The results of this model and prototype study will be presented in Section 5.6. Finally, Section 5.7 will conclude that a bisecting hip mechanism indeed provides an elegantly simple solution for stable and efficient walking with an upper body. The content of this chapter is based on two of our papers [183, 180], © 2007 IEEE, reprinted with permission.

5.2 Point Mass Model with Upper Body

The goal of this chapter is to add an upper body to an (almost) passive walking robot. We start with a simplified model, as earlier described in [183]. This model should be as simple as possible for the sake of a minimal set of parameters, so a natural starting point was the 'simplest walking model' of Garcia *et al.* [54], see Chapter 2. The simplest walking model consists of two rigid massless legs, with small pointmasses m_f as feet and a finite pointmass at the frictionless hip joint. For slopes up to 0.015 rad, this model performs a stable walk downhill.

This simple model is extended with an accordingly simple upper body. A point mass is connected to a rigid, massless stick that rotates around the hip joint (Fig. 5.2). The upper body is parameterized with body length l_b and body mass m_b. The parameters have been given default values that have some relevance to human walking or to future prototypes, see Table 5.1. We made the parameter values dimensionless for comparison with other models: all sizes are scaled with the leg length, so that the leg length is 1, and all masses are scaled with the sum of the pelvis mass and the upper body mass, so that the pelvis mass is $1 - m_b$. The foot mass is not included in this sum for reasons of compatibility with older models [141]. Time is scaled so that the resulting gravity is 1. There are also two non-human parameters: 1) slope angle γ, by which we can tune the walking speed, and 2) hip spring stiffness k, which allows tuning of the step frequency. The spring will turn out to be necessary for stable walking. With the default parameter values according to Table 5.1, the model walks with human-like speed and step length.

Fig. 5.1. Prototype 'Max'; a 2D passive dynamic walking robot with an upper body connected to a bisecting mechanism at the hip. Movies of this prototype in action can be found at our website [1].

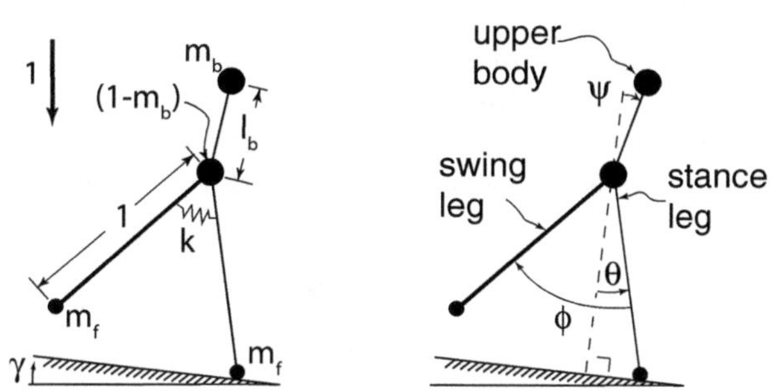

Fig. 5.2. Model of the simplest walker with upper body; **Left:** parameters. **Right:** degrees of freedom.

As described so far, the model would have three degrees of freedom (Fig. 5.2): absolute stance leg angle θ (counter-clockwise), relative swing leg angle ϕ (clockwise), and absolute body angle ψ (clockwise). However, the upper body would

Table 5.1. Default parameter values for the simplest walker with upper body, from a rough estimation of human proportions. The parameters are nondimensionalized by scaling: mass is divided by ($\overline{\textit{pelvis mass} + \textit{upper body mass}}$), length is divided by *leg length*, time is divided by $\sqrt{\textit{leg length}/\textit{gravity}}$.

parameter	symbol	approximated human value	scaled
Foot mass	m_f	7 kg	0.1
Upper body mass	m_b	49 kg	0.7
(Pelvis mass)	$(1 - m_b)$	21 kg	0.3
Leg length	-	1 m	1
Body length	l_b	0.4 m	0.4
Hip spring stiffness	k		0.4
Slope angle	γ		0.0725 rad

then just be an inverted pendulum jointed around the hip. Without any active control acting on it, one can expect that it will not be kept upright passively. To keep a fully passive upper body upright, A. Ruina (personal communication) suggested four possibilities:

1. Use a light upper body that has its actual center of mass *below* the hip. This option is not very useful in realistic prototypes.
2. Use springs that keep the upper body upright [160]: This also has the utility that it should give more efficient walking by making the steps smaller at a given speed [97].
3. Use a bisecting mechanism: a kinematic coupling that keeps the body midway between the two legs (Fig. 5.3).
4. Keep the model as is, and hope that for some special mass distribution suddenly a stable gait emerges.

Intuitively, option three is most promising because the number of degrees of freedom is reduced, which improves the chances of finding stable walking cycles. Human beings do not have such a kinematic coupling, but the assembly of pelvic muscles and reflexes could possibly perform a similar function. Also,

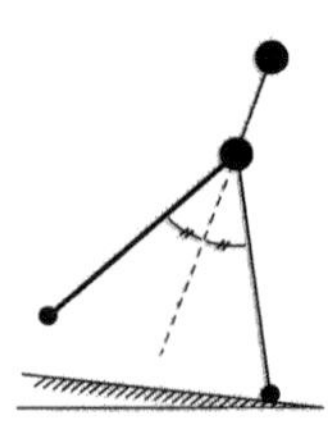

Fig. 5.3. Kinematic coupling of the upper body to the midway leg angle according to Eq. (5.1)

such a construction can be found in certain reciprocating gait orthoses [81]. In robot prototypes such a kinematic coupling can be easily realized. In the model it is introduced according to:

$$\psi = \phi/2 - \theta. \tag{5.1}$$

The other options could provide valuable results, although the first is not interesting as a model for human walking. We intend to investigate options two and four in the future, but in this book we will focus on the behavior of the model with the compass-like kinematic constraint.

5.3 Results of Point Mass Model

5.3.1 Walking Motion

The walking motion is analyzed with the help of the methods as described in Chapter 2. With the default parameter values, the model takes something like a human walking step if started with the initial conditions from Table 5.2. However, due to its quintessential nature our model shares some typical non-human characteristics with the simplest walking model of Garcia *et al.* [54]. First, the feet are no more than points, hence the application point of the ground contact force is at a fixed location during one step. Second, there are no actuators, so that the model will only walk if placed on a slope. Third, the legs cannot change length, hence there are not enough degrees of freedom to allow a double support phase.

Table 5.2. Initial conditions that result in a cyclic walking pattern for the simplest walker with upper body, using the default parameter values (Table 5.1)

θ_0	0.3821 rad
$(\phi_0 = 2\theta_0)$	(0.7642 rad)
$\dot{\theta}_0$	-0.3535 rad/s
$\dot{\phi}_0$	0.0736 rad/s

The step starts and ends immediately after 'heel strike' (Fig. 5.4). The hip moves forward like an inverted pendulum with an almost constant speed, while at the same time the swing leg swings to a forward position. Naturally, the kinematic constraint keeps the upper body at the bisecting angle midway between the two legs. The motion of the swing leg appears to be that of a free pendulum, while actually it is mainly the result of the dynamics of the upper body and the hip spring.

The trajectories of the various point masses are no surprise; the hip moves forward on a circular path (often referred to as 'compass gait' [83]), while the swing foot remains close to the floor. The upper body follows a path almost identical to the hip trajectory at a distance l_b above the hip, only slightly smoother at the heel strike discontinuities. There are two peculiarities. First, the hip trajectory equals that of an inverted pendulum, but its speed does not. Due to the influence of the upper body and the hip spring, the speed of the hip is nearly constant, as can be deducted from the nearly constant stance leg velocity in Fig. 5.4. Second, the swing foot travels briefly below floor level. Inevitable for a 2D walker with straight legs, we allow this to happen in our simulation. Human beings and our more sophisticated models and prototypes have knees to solve this problem.

With a scaled step length of 0.746 and a scaled step time of 1.77 the model attains a scaled walking velocity of 0.42. Back on a human scale (on earth) this corresponds to 1.3m/s. The scaled velocity is the same as the familiar Froude number, $\sqrt{v^2/gl}$, where {Froude number = 1} represents the maximum walking speed for any biped. At higher speeds the foot contact force would become negative, so the biped should switch to running or maybe to Groucho walking. With a Froude number of 0.42 our model is well below that boundary, firmly stepping its way but not even close to running.

The energy consumption of the model at this speed is low. This is usually [35] represented in the non-dimensional 'Cost of Transport' c:

$$c = \frac{E}{mgs} \tag{5.2}$$

i.e. the energy consumption E per distance traveled s per unit of weight (= mass m times gravity g). For passive dynamic walkers the specific resistance is equal to the sine of the slope angle $\sin\gamma$ as gravity is the only means of energy input. So, our model has a specific resistance of 0.0725 at a (scaled) speed of 0.42. This is much more efficient than human beings walking at the same speed with a specific resistance of approximately 0.38 [133], although the comparison is somewhat unfair as muscle efficiency is unaccounted for. Also, this is much more efficient than most other walking robots [35].

5.3.2 Inherent Stability

To classify the stability of the walking motion there are two useful but essentially different definitions. First, we can regard stability in its most strict way. The basis is a walking motion in cyclic equilibrium, called a 'limit cycle'; a certain combination of initial conditions (Table 5.2) keeps repeating itself for all subsequent steps. If started slightly away from the limit cycle, the walking motion is stable if the subsequent step is closer to the limit cycle. Note that this 'local stability' requires the existence of a limit cycle, and that only small disturbances are investigated. We found that the model with the parameter values from Table 5.1 and started with the initial conditions from Table 5.2 is indeed stable for small disturbances. The cycle has a Monodromy matrix with eigenvalues ("Floquet Multipliers", i.e. error multiplication factors) that all lie within

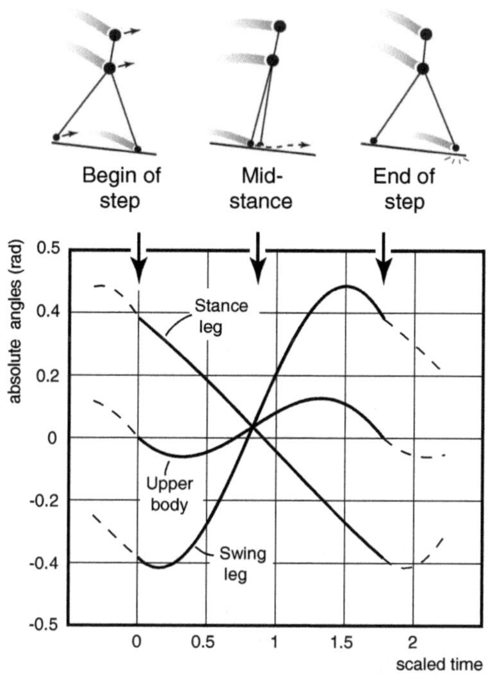

Fig. 5.4. Cyclic walking motion of the model with upper body. Top: stick figure representation, bottom: absolute angles of stance leg, swing leg, and upper body. The simulation is performed using the default parameter values (Table 5.1).

the unit circle, which means that small errors decay step after step and thus the cycle is stable.

Second, we can regard the stability of walking in the broadest and most intuitive form: 'The robot is stable if it does not fall'. We can even allow ourselves to use the formally incorrect term *'more stable'* for a robot that can handle larger disturbances. Note that this 'global stability' does not require the existence of a limit cycle (every step may be different, as long as the robot doesn't fall), but that it can only be investigated with the costly method of trying out all possible disturbances.

By application of the cell mapping method [78], we found that the model performs surprisingly well. The model converges to its limit cycle if started with errors as large as 8% on all initial conditions of Table 5.2, compared to 2% for the simplest walking model [141]. For certain combinations of errors, the errors can even be much larger. This is inspected by evaluation of the *basin of attraction* (Fig. 5.5), the complete set of initial conditions that eventually lead to cyclic walking. For example, the figure shows that cyclic walking with cyclic initial conditions as in Table 5.2 emerges even if the initial step is twice as large, e.g. $\{\theta_0 = 0.75, \dot{\theta}_0 = -0.75, \dot{\phi}_0 = -1\}$.

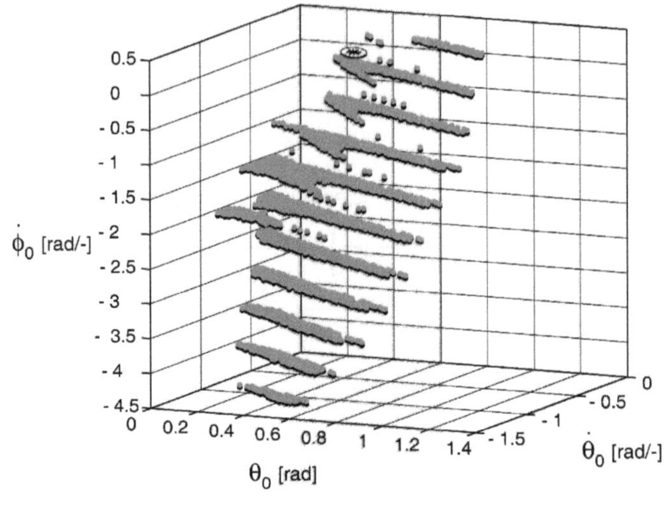

Fig. 5.5. Basin of attraction of the simplest walking model with upper body. The gray layers of points represent horizontal slices of a 3D region of initial conditions that eventually lead to the cyclic walking motion. The cyclic motion ($\{\theta_0 = 0.3821, \dot{\theta}_0 = -0.3535, \dot{\phi}_0 = 0.0736\}$, Table 5.2) is indicated with a flat asterisk on a small disk, just above one of the sample slices.

5.4 Parameter Study on Point Mass Model

5.4.1 Slope and Spring Stiffness; Speed and Step Length

As mentioned in Section 5.2, the model has two parameters that are essential to the model's gait: the slope angle and the hip spring stiffness. Together, they determine the step frequency and the step length, thereby also determining the walking velocity.

First, for a fixed set of mass and length parameters, the step frequency is mainly determined by the hip spring stiffness. It appears that the swing leg amplitude, step length, slope angle or walking speed all have a negligible influence on the step frequency [108, 54].

Then, the step length is directly determined by the slope angle; the steeper the slope, the larger the steps. This is a result of the balance between the gravitational energy input and the impulsive energy losses at heel strike. Although a larger step means more energy input, it leads to even more energy loss at heel strike. As a result, the system will automatically converge to a periodic walking motion with a step length that corresponds to the slope angle.

With the hip spring stiffness and the slope angle together, we were able to set both the speed and the step length to human values. It should be noted that these effects are not unique to our model. In fact, Kuo [97] studied these same effects extensively for the simplest walking model to investigate energy matters of human walking.

5.4.2 Upper Body Height and Weight

The upper body is parameterized with body length l_b and body mass m_b. The body mass and the pointmass at the pelvis together always sum to 1 for the purpose of scaling, while the body length is scaled to the length of the leg. The default parameters of Table 5.1 are chosen so that the model has some relevance to future prototypes. This section investigates the model behavior when the upper body is reduced to nothing or significantly enlarged.

Reduction of the upper body size or mass to zero leads to a model like the simplest walker, except that the simplest walker has no hip spring and an infinitesimally small foot mass. For a very small foot mass, no hip spring is necessary, but for a realistic foot mass as in Table 5.1, stable walking cycles only exist if a spring is applied. As stated earlier, the hip spring and slope angle together determine the walking speed and the step length. If we set them so that speed and step length match the original model (Table 5.1), we find that the 'zero-body-model' needs a slope angle of $\gamma = 0.147\ rad$. In other words, the model with upper body is twice as efficient as the same model without upper body! Apart from that, there is not much difference between the gaits of the two models.

Similarly, an increase in the mass or the size of the upper body will provide an even higher walking efficiency. We found that an increase in m_b has a similar effect as an increase in l_b. As an example, we crudely modeled a person carrying a heavy load on the top of the head by setting $m_b = 0.9$ and $l_b = 1$. The hip spring stiffness and slope angle were again adjusted to obtain human walking speed and step length. The required slope angle is now only $\gamma = 0.0249\ rad$; this model walks about three times more efficient than with the default parameter values! In general it is clear that the presence of an upper body has a positive influence on the walking efficiency.

The changes of the mass or size of the upper body have little effect on the stability. We investigated the three previously mentioned situations: a) zero upper body mass, b) default parameters (Table 5.1), and c) someone carrying a heavy load on the head ($l_b = 1, m_b = 0.9$). In terms of linearized stability, all three situations are stable for small disturbances. In terms of global stability, the allowable errors on all initial conditions are about 8% for all three situations. However, the resultant basins of attraction (as in Fig. 5.5) have different shapes, so that convergence from larger errors occurs for different combinations of errors. It seems odd that the size or mass of the upper body has no apparent influence on the allowable errors (all 8%), whereas there is such a large difference with the simplest walking model (only 2%). We believe that this is a result of the increased speed and step frequency; the simplest walking model walks slower than our model, which we tuned to walk with human speed. We intend to investigate this effect in the near future.

5.4.3 Limits to Stability

Our upper-body walker has a remarkably stable gait if provided with the parameter values from Table 5.1. For certain other parameter values, however, the model has unstable gaits or even no cyclic walking motions at all. Usually this can be solved by sufficiently increasing the hip spring stiffness k, with a few exceptions. At slopes steeper than $\gamma \approx 0.35$ rad, the equilibrium speed is so high that the stance foot would loose ground contact and the model should start running. The foot mass m_f and the body size and mass l_b and m_b can be chosen arbitrarily small or large; with a sufficiently high value for k the model still walks fine, although this could result in correspondingly small or large step lengths, which in turn could lead to the loss of floor contact.

Inside these boundaries, for each combination of parameter values there exists a minimal value for k that ensures stability. For the model with the default parameter values of Table 5.1, we studied the effect of variations in k on the cyclic walking motion. For $k > 0.218$ we found steady, stable cyclic walking. However, for the same value of k there also exists a second, unstable gait. The steps are shorter and faster, and the motion looks like the model is stumbling forward. McGeer [108] and Garcia *et al.* [54] discovered this second solution for their models and refer to it as the 'short-period gait', as opposed to the normal, stable solution which is termed 'long-period gait'. We are only interested in the last type of gait, the behavior of which we have studied as a function of the parameter value for k.

Above the boundary value, an increase in k results in faster and smaller steps. If we decrease k below 0.218, we cross a bifurcation to asymmetric gaits, first encountering period-two solutions and for lower k even higher-period solutions. These solutions are still stable. Below $k = 0.162$, we found only unstable gaits or even no cyclic solutions at all. Garcia *et al.* [54] found a similar bifurcation to chaos for the simplest walking model when increasing the slope above $\gamma = 0.015$ rad.

We tracked the first bifurcation point over a range of parameter values because that point represents the minimally required value for k to obtain normal, stable walking. The relation between the minimal value for k and the other parameters is not linear, and there is not an obvious and simple non-linear relationship. Qualitatively, the required hip spring stiffness k needs to be increased if l_b, m_b, m_f or γ are increased.

5.5 Prototype Design

5.5.1 Simulation Model

We now move from the point mass model to a realistic model of a two-dimensional 5-link biped (Fig. 5.6). The model has a common topology; the upper body is a single rigid link, while each leg consists of a thigh and a shank interconnected through a knee joint. The knees are equipped with a hyperextension stop (assuming fully inelastic impacts) and a locking mechanism (latch) which is released just after the start of the swing phase. With the bisecting hip mechanism, the total number of degrees of freedom is at most 3; absolute upper body angle ψ, inter-leg angle ϕ, and relative swing knee angle ξ. At the end of a step when the swing knee is fully extended, only two independent degrees of freedom remain (four states; two angles and their velocities). Note that this is the same number of degrees of freedom as for kneed walkers without an upper body (Chapter 4) due to the constraint of the bisecting hip joint.

Ankle joints are not present, as rigidly attached arc feet have proven to be a simple and sufficient solution for stable passive walking. We assume that the links suffer no flexible deformation and that the joints are free of damping or friction. Also, we assume a perfect bisecting mechanical coupling between the legs and the upper body. The contact between the foot and the floor is idealized, assuming perfectly circular feet that do not deform or slip, while the heel strike impact is modeled as an instantaneous, fully inelastic impact where no slip and no bounce occurs. The walker walks on level ground and thus requires a small amount of energy input per step. This is provided by means of the hip muscles which accelerate the swing leg to a forward position. Their main function is to provide fore-aft stability, but their secondary effect is the input of just enough energy into the system to maintain the cyclic walking motion. Finally, the floor is assumed to be a rigid, flat, and level surface.

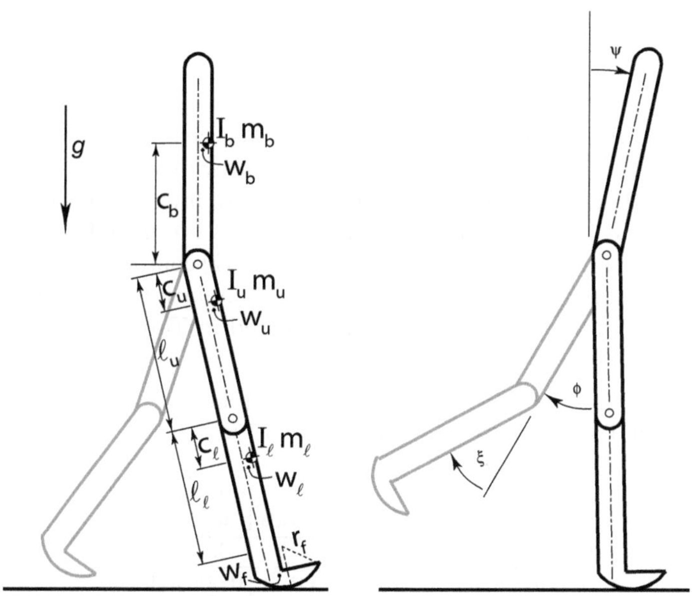

Fig. 5.6. Two-dimensional 5-link model. **Left:** parameter definitions. **Right:** degrees of freedom (ψ is not measurable in the prototype).

5.5.2 Simulation Procedure

The simulation procedure is similar to that applied in the point mass model study of the previous sections. The procedure is a succession of nonlinear numerical dynamic simulations of walking steps which begin and end at the instant immediately after heel strike. Within one step, the equations of motion are numerically integrated until an event is detected such as knee strike or heel strike, followed by an impact calculation. After the heel strike impact the simulation of the walking step is ended. The end state of the walker (an instantaneous double stance phase) can then be used as the starting state for the next step, or it can be compared to the initial state of the walker. If the end state equals the initial state, we have found a fixed point representing a cyclic walking motion. We then apply the Poincaré Mapping method for stability analysis. Additionally, to investigate 'how stable' the walking motion is, we perform an approximate search for the boundaries of the basin of attraction of the fixed point. For this realistic model, it is unfeasible to do a full investigation of the basin of attraction, so we cannot apply the cell mapping method as was done with the point mass model. Instead, a walking step is simulated with initial conditions that deviate from the fixed point in eight different combinations of states (e.g. a positive deviation on the stance leg angle combined with a negative deviation on the angular velocity of the body). We search for the largest allowable deviations that still lead to successful walking. The resulting estimate for the boundaries of the basin of attraction are a measure for the size of disturbances (at the start of a step) that the walker can still recover from.

5.5.3 Default Parameter Values

A set of physically realistic parameter values that lead to stable walking was readily found. Re-using partial designs from the previous prototype (Chapter 4) we arrived at a 10 kg machine with a 0.6 m leg length and 1.1 m total height. The physical properties, such as the mass distribution were initially determined by convenient placement of the electronic and pneumatic components. The resultant configuration resulted in stable walking in the simulation, so we have adopted these parameter values as the default values listed in Table 5.3.

Table 5.3. Default parameter values for the prototype with two full CO_2 canisters. The underlined u, l, and b serve as subscribts as in Fig. 5.6.

	body	upper leg	lower leg
mass m [kg]	8	0.7	0.7
mom. of Inertia I [kgm^2]	0.11	0.005	0.005
length l [m]	0.45	0.3	0.33
vert. dist. CoM c [m]	0.2	0.15	0.16
hor. offset CoM w [m]	0	0	0
foot radius r_f [m]	-	-	0.25
foot hor. offset w_f [m]	-	-	0.01

5.5.4 Construction of the Prototype

The central part of the prototype (Fig. 5.1) is its bisecting hip mechanism. Of the many possible forms of implementation we chose to apply an auxiliary axle connected to the legs with one straight and one cross-over chain (Fig. 5.7). In hindsight, it is valuable to report that this solution requires extra attention to the problem of slack in the chains. Also, one must be aware that rather large torques are transmitted through the chainwheels and axles, especially when the prototype occasionally falls. Nonetheless, for our relatively lightweight prototype this solution is satisfactory. Other possible mechanisms include a four-bar linkage [34], a differential gearbox, or cables and pulleys (as applied in some gait orthoses [81]). Alternatively, the bisecting hip action can also be obtained in fully actuated robots where a local controller keeps the upper body in the bisection angle [130, 25].

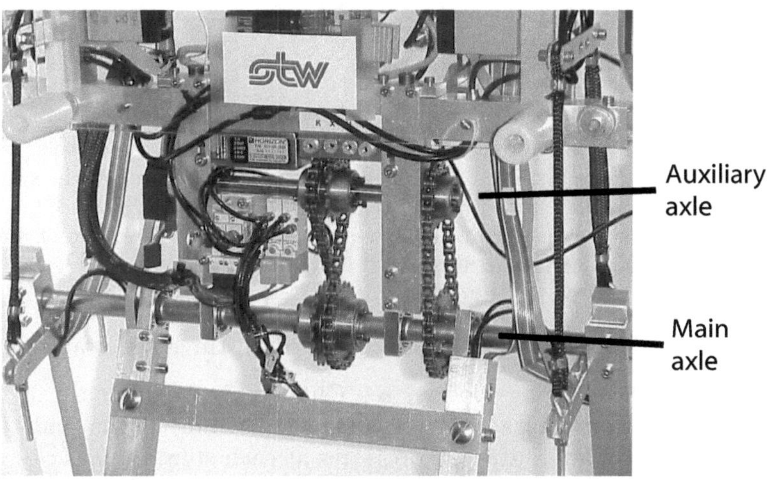

Fig. 5.7. The bisecting hip mechanism in the prototype. The outer legs are rigidly attached to the hip axle, the inner legs can rotate freely. The hip axle is connected through bicycle chains via an auxiliary axle to the inner legs.

The prototype is autonomously powered with an on-board pneumatic system. The pneumatic components are displayed in Fig. 5.8, clockwise arranged according to the CO_2 flow through the system. The returnable Alco$_2$jet^{TM} canister (widely available for home soda machines) contains 450 g CO_2 at the saturation pressure of 5.8 MPa and weighs 1.2 kg when completely full. The pressure is reduced in two stages, first to approximately 1.2 ± 0.2 MPa and then to 0.6 ± 0.01 MPa. Both levels are manually adjustable. We developed the regulators specially for this project because they are not commercially available in the required small and lightweight design (the small 40x20x10 mm block in Fig. 5.8 actually contains *four* second stage regulators). The second stage pressure output is fed via low-power SMC^{TM} valves to four tiny SMC^{TM} cylinders that control the knee latches and to four Shadow^{TM} McKibben muscles that act as two antagonistic pairs between the robot's body and the outer legs (attached with a moment arm of 60 mm).

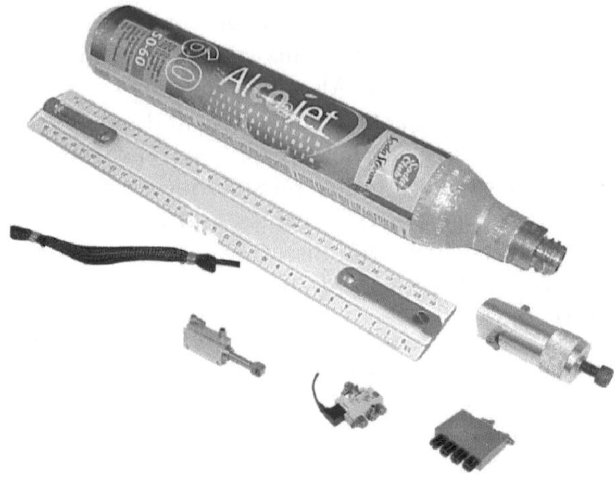

Fig. 5.8. The pneumatic components and a 30 cm ruler to indicate their sizes. The components are clockwise arranged according to the CO_2 flow through the system; 5.8 Mpa CO_2 canister, the first stage pressure regulator to 1.2 Mpa, a block of four second stage regulators to 0.6 Mpa, one of four low-power SMC^{TM} valves, one of four small SMC^{TM} cylinders and one of four Shadow^{TM} McKibben muscles.

The McKibben muscles in this prototype perform three simultaneous tasks:

1. They power the walking motion, see Chapter 3. A difference of internal pressure between two antagonists results in an asymmetry that pulls forward the swing leg. By alternation of pressures at each step, the muscles inject a small amount of energy into the system and thus replenishes the energy lost in damping and impacts.
2. They provide robustness against falling forward, see Chapter 4. Especially the non-linear behavior near maximal extension is beneficial for this, as the muscles effectively slow down the forward-rushing swing leg and then keep it in that forward position.
3. They provide the required hip spring stiffness for the upper body (this Chapter).

The control system is exactly the same as in the previous prototype (Chapter 4). There is one foot switch underneath the most right foot, and one underneath one of the middle feet. These two switches are read by a micro-controller, which then triggers only two valve actions per step based on the state of the foot switches. If the inner leg's switch is contacted, the front hip muscles are switched to high pressure and the antagonists to low pressure, effectively pulling the outer legs forward. Simultaneously, the knee latches of the outer legs are released briefly. Then, the system just waits for the outer leg's foot switch to make contact, assuming that knee extension takes place before heel contact. The entire control algorithm is easily implemented in any microcontroller (we have experimented both with a Microchip™ PIC16f877 and with a LEGO Mindstorms™ RCX controller).

For post-experiment data analysis, however, a more elaborate electronic system is required. The prototype is equipped with four optical encoders (hip, inner knee, left and right outer knee) and with one gyroscope mounted in the robot's body. The low-level processing (counters and A/D conversion) is still done in a PIC microcontroller, while the data is collected at 50 Hz in a J-stick™ Java board that can be read out after the experiments. Even with the measurement system active, the entire robot remains fully autonomous.

5.6 Prototype Experiments

5.6.1 Resultant Motion and Gait Characteristics

The gait of the prototype looks natural, see video at [1]. The resultant walking motion is depicted in Figure 5.9, in which we have plotted both the simulation results and the actual prototype recordings. The figure presents the absolute body angle (simulation only, not measured in the prototype), the relative hip angle and the knee angles as a function of time, together with the foot clearance (also simulation only). The clearance amounts to 5 mm or more throughout the step. The body remains approximately upright with maximal excursions of $\pm$ 0.15 rad. The knee reaches full extension 0.5 s after the start of the swing phase. The maximal inter-leg angle is $\pm$ 0.65 rad, but at the time of heel strike this is $\pm$ 0.55 rad, leading to a step length of 0.35 m. The model is walking in its limit cycle, taking 1.2 steps per second thus walking at 0.42 m/s (Table 5.4).

Table 5.4. Gait characteristics when walking with the default parameter values

Step length	0.35 m
Step frequency	1.2 Hz
Velocity	0.42 m/s
Nominal clearance	5 mm
Cost of Transport	0.2 $J/(kgm^2s^{-2})$

The differences between the motions of the model and the prototype are small, especially when considering that the model is walking in its limit cycle while the prototype is only *close to* its limit cycle due to constant disturbances; the floor is far from perfectly flat and level. A noticeable difference is in the amount of knee flexion. Especially the knees of the outer legs bend less than predicted by the simulation, probably caused by friction and damping in the knee joint or by a slight delay in the knee latches. The overall effect on the walking motion is small, except for the foot clearance which then decreases significantly and indeed causes most of the failures.

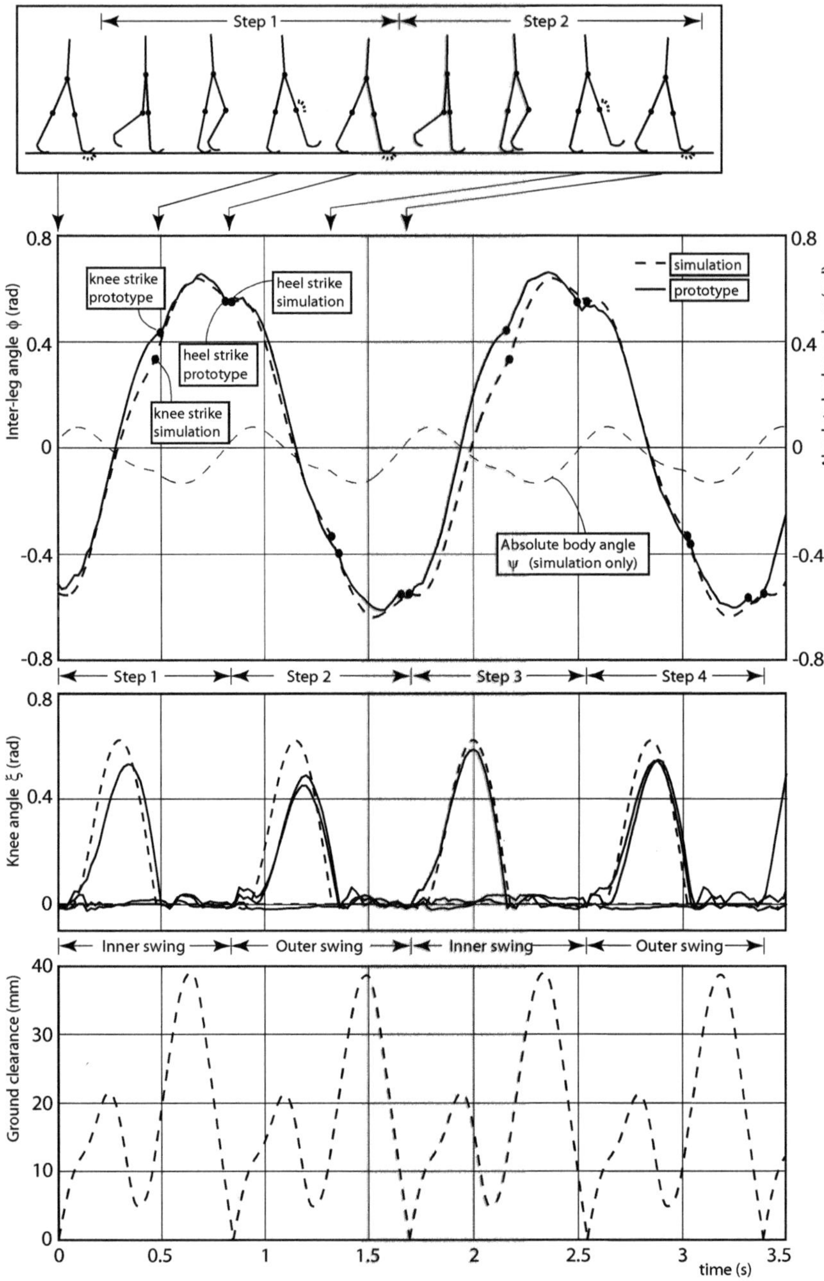

Fig. 5.9. Comparison of the walking motion of the simulation (dashed lines) and the prototype (solid lines). The absolute body angle ψ and the clearance were not measured in the prototype. The inter-leg angle shows a slight asymmetry in the prototype's gait. The knees of the prototype show approximately 0.05 rad play of the latch.

5.6.2 Stability

The stability of the cyclic walking motion is usually analyzed by investigating the initial states of each step in a sustained walking motion. In this chapter, we choose to depart from the usual approach and to investigate the *end* states instead of the initial states. The difference is that the analysis here is based on the velocities just *before* heel strike in contrast to the tradition of using the velocities just *after* heel strike. The reason is that the velocity measurements in the prototype are unreliable just after an impact due to transient oscillations in the mechanical system.

At the end of a step, with both feet simultaneously on the floor and with both knees extended, there are only three independent states; inter-leg angle ϕ, its angular velocity $\dot{\phi}$ and the absolute angular velocity of the body $\dot{\psi}$. Their fixed point values given in Table 5.5 are determined with the computer simulation for the parameter values in Table 5.3. A Floquet Multiplier analysis of the fixed point on the computer model predicts that the walking motion is stable, i.e. that small errors on the end states in Table 5.5 decay step after step.

Table 5.5. Fixed point values for the three independent end states just before heel strike, valid for the parameter values from Table 5.3. The difference between simulation and prototype arises from the simplified model for the non-linear muscle behavior.

		simulation	prototype (average ± standard deviation)
Inter-leg angle	ϕ rad	0.55	0.55 ±0.06
Inter-leg ang. vel.	$\dot{\phi}$ rad/s	-0.58	-1.15 ±0.61
Upper body ang. vel.	$\dot{\psi}$ rad/s	1.04	1.03 ±0.17

Fig. 5.10 shows the walking results of over 200 steps (measured in series of 40 steps on average) depicted in the phase plane. The graph only shows two out of the three independent states because the inter-leg angular velocity $\dot{\phi}$ is not relevant; Table 5.5 shows a high variability for this state and the simulations have shown us that even much larger variations on this state can be allowed without resulting in a failure. The reason for this insensitivity is the fact that the inter-leg angle is controlled by the hip muscles toward a fixed end position, independent of the initial velocity. The difference between the measured average and the simulated value for $\dot{\phi}$ is a direct result of the simplified model for the muscle non-linearities at maximal extension.

The experimental results are indicated with black dots in Fig. 5.10. The last step in a series is indicated with an encircled cross, because it is the last step before a fall. These experimental results correspond well with the simulation results, which are indicated in the figure with the gray area. According to the simulation model, the gray area is the basin of attraction; a start outside the area will either lead to a fall Forward or a fall Backward. The average state

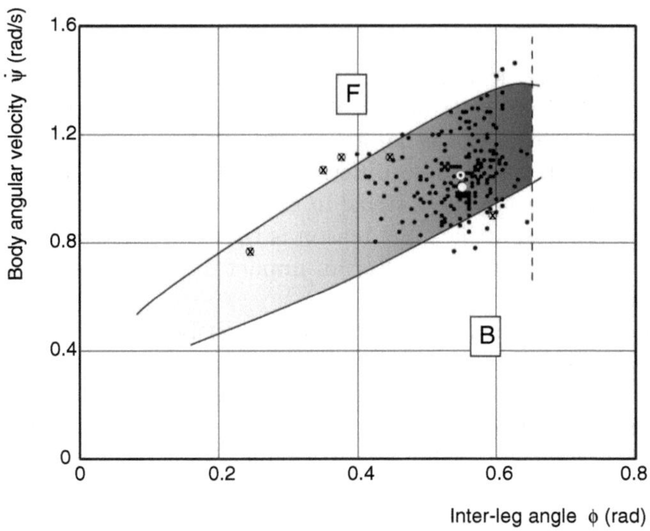

Fig. 5.10. A section of the basin of attraction in the Poincaré map. The figure shows the two most sensitive states at the end of a step, namely the inter-leg angle ϕ and the angular velocity of the upper body $\dot{\psi}$. The walker is not sensitive to variations in $\dot{\phi}$, the third independent end state, which is therefore not shown. The black dots represent 200 measured states during continuous walking, whereas the last step of each series of steps (the last before a fall) is indicated with an encircled 'x'. The boundaries of the basin of attraction as derived from the simulation are given by the solid black lines. Below the lower boundary, the robot falls Backward, above the upper boundary it falls Forward as a result of foot-scuffing. The dashed line represents maximal extension of the hip muscles. Due to the hip actuation the robot is not likely to arrive in the lower left part of the basin of attraction, but if it would, it would return stably to its limit cycle. The fixed point of the simulation is a white circle, the average measured state is a white circle with a dot.

(indicated in Fig. 5.10 with a white encircled dot) also corresponds neatly to the fixed point from the simulation model (white circle), see also Table 5.5.

The stability results indicate that the prototype can be easily started with a manual launch (illustrated in Fig. 5.11). Moreover, we could also realize an automatic launch from a static position, although this only works if the legs are placed with a very small inter-leg-angle, i.e. with the four legs almost all parallel. After a launch, the prototype can walk indefinitely on a level floor until it runs out of power or into a wall. In contrast to the robustness against disturbances in a manual launch, the walker appears to be not too robust against variations in height in the floor surface. The variability of the measurements is quite large. This is a result from the irregularities in the hallway floor. The floor has variations in height of maximally 3.5 mm in one step, amounting to a local slope of $\pm$ 0.5°. These irregularities are close to the maximal allowable disturbances as predicted by the simulation model, explaining why some of the measurement points are close to the boundaries of the basin of attraction. The simulation

Fig. 5.11. Video stills illustrating the walking motion after a manual launch. Video is available at our website [1].

predicts that the walker can handle a step down in the floor of maximally 3 mm. We verified this with an experimental setup where it walked on a rigid, flat and level surface (not the hallway floor) and then took a step down. Indeed it could handle not much more than 3 mm.

5.6.3 Parameter Sensitivity

The prototype is tolerant to variations in most of the parameters (e.g. 1 kg of extra mass on the upper body has no noticeable effect), except for those parameters that affect the forward velocity. The forward velocity is the net result of the velocity increase during the stance phase and the instantaneous velocity decrease at heel strike. The velocity increase is determined by the amount of time that the robot's center of mass spends *behind* the foot contact point (deceleration) and the amount of time spent *in front of* the contact point (acceleration). Any parameter that influences these has a strong effect on the walking motion; with too much deceleration the walker will have a tendency to fall backward whereas with too much acceleration the resultant walking velocity will be high and thus the chances of falling forward increase.

Parameters with a direct influence are w_b, w_u, and w_l (Fig. 5.6 and Table 5.3) which determine the horizontal position of the center of mass, and w_f and r_f which determine the foot contact point. The effect of the position of the center of mass is strong. For our 10 kg walker, a 500 g additional mass that can be attached up to 100 mm in front or behind the hip joint already provides sufficient tuning possibilities. In our opinion, the automatic control of the fore-aft balance will be one of the major improvements for future dynamic walking robots.

The foot radius r_f determines how much the foot contact point travels forward during the stance phase and thus a larger radius has a weakening effect on both the robot's deceleration and acceleration. Previous experiments and simulations [186] have shown that this effect is beneficial to the robot's robustness against disturbances. A forward foot offset $w_f > 0$ creates a forward tilt of the entire robot (best visualized in a drawing of the heel strike state) and thus results in faster walking. Therefore, an increase in w_f should be accompanied by a backward displacement of the center of mass. Note that this observation is only valid for a walker with an upper body with the bisecting hip mechanism and with a substantial mass at a substantial distance above the hip joint. For walkers without an upper body, the effect is reversed.

The hip actuation has an indirect but significant influence on the deceleration and acceleration during the stance phase. For any walker with physically feasible parameter values (also without upper body), the center of mass moves forward when the swing leg is swung forward. This is best verified in a simplified analysis without gravity. If the swing leg is moved quickly by a strong hip actuation, then that forward displacement takes place early in the stance phase, and thus the center of mass will spend relatively more time in front of the foot contact point. In other words, the faster the swing leg is moved forward, the faster the robot will walk. The strength of this effect depends on the amount of inertia (of both the legs and the upper body) that is involved when the hip actuators are engaged.

There seems to be a counterproductive effect here, as the hip actuation was installed in the first place to *reduce* the chances of falling forward and now it appears to *increase* that chance by increasing the walking velocity. This can be resolved easily, however, with a backward adjustment of the robot's center of mass so that the total effect (of hip actuation and mass displacement) is an enlargement of the basin of attraction.

In conclusion, the parameters of the upper body barely influence the walking behavior and the stability. There is almost no effect of an increase of the mass or a vertical displacement of the center of mass. Only the fore-aft position of the center of mass is important, as it regulates the average forward walking velocity, and thus the chances of falling forward or backward.

5.6.4 Energy Efficiency

The specific cost of transport (Eq. 5.2) of our prototype is calculated with the CO_2 expansion through the muscles from the 0.6 MPa input pressure to 0.24 MPa relief pressure. The prototype uses 208 mg CO_2 per step (allowing it to walk for 30 minutes on a single canister). The exergy (or 'availability'), i.e. the amount of work that could theoretically be done with gas expanding from 0.6 MPa to 0.24 MPa, is 10.6 Joule per step, so the specific cost of transport equals 0.32. Although the specific cost of transport for the prototype resembles that of a walking human being, some deliberations must be taken into account.

On the one hand, one could argue that the prototype is much more efficient than the human. The pneumatic muscles are not optimal for their task, because they have a fairly large 'dead volume' which must pressurized at each action cycle. They use much more pneumatic energy than the amount of work they produce. We determined with the simulation model that the amount of work produced by the muscles (i.e. their force integrated over their elongation) is only 0.5 Joule per step, leading to a very low specific cost of transport of 0.01. Note that this value is in the same range of the fully passive walkers of Chapter 2.

On the other hand, one could argue that the prototype is much less efficient than the human. The specific cost of transport for the human includes the metabolic cost of the entire system, and specifies how well the available energy is used. In that respect, it would be fairer for the prototype calculations to also include the idle pressure reduction from 5.8 MPa to 0.6 MPa. Although exact figures are not available, it is certain that the total amount of available pneumatic energy from the CO_2 canister is factors higher than the energy that is used in the muscles. However, the main cause of this apparent waste of available energy lies not in the applied concept of passive dynamic walking but rather in the unavailability of pneumatic components that can use the energy of the high-pressure canister. It is expected that ongoing research in the field of pneumatics will eventually solve this problem.

5.7 Conclusion

This chapter reports on the successful addition of an upper body to a walking robot based on the concept of passive dynamic walking. The upper body is connected to the legs by means of a bisecting hip mechanism which forms a passive solution to stabilize the upper body while simultaneously allowing a passive swing leg motion. The prototype walks stably and efficiently. The fore-aft position of the center of mass of the upper body is a powerful parameter for the stability of the walking motion. Conversely, the height of the center of mass, the total mass and the mass distribution have no noticeable influence on the performance. Thus, we conclude that the bisecting hip mechanism forms a practical and simple solution to construct efficient bipedal walking robots, in agreement with the concept of passive dynamic walking.

The simulation results suggest that the capability to reject larger disturbances increases when the model walks faster. So far, this effect seems to be unrelated to the added upper body. It is an interesting effect that we intend to research in the near future.

6. Denise; Sideways Stability

M. Wisse and R.Q. van der Linde: Delft Pneumatic Bipeds, STAR 34, pp. 85–115, 2007.
springerlink.com

6.1 Introduction

One of the grand challenges in biped walking is to find the key to stability in three dimensions. In addition to the fore-aft motions (pitch), in 3D also sideways motions (lean) and rotations around the vertical axis (yaw) are possible. It is the interaction between all three of these that makes the problem of 3D stability so difficult. Consequently, most of the known solutions to this problem are successful because they *reduce the interaction* in one way or another:

- Ignore yaw by assuming sufficient yaw resistance in the foot contact [96, 127].
- Apply large moments of inertia against yaw [33, 32].
- Walk with short steps [160, 33, 150].
- Counteract yaw with a counter-rotating body [181] or with counter-swinging arms [36].

In this chapter, however, we aim not at a *reduced interaction* between the degrees of freedom, but conversely we show how a *purposefully induced interaction* between lean and yaw can actually improve the stability of the walking motion.

First, Section 6.2 introduces the main idea through the related problems of skateboard and bike stability. Then in Section 6.3 an elementary walking model is introduced, followed by the simulation results in Section 6.4. Sections 6.5 to 6.7 discuss the application of the idea in a prototype that was developed for this purpose. The content of this chapter is based on three of our papers [182, 176, 178], © 2004, 2007 IEEE, reprinted with permission.

6.2 Advantageous Lean-to-Yaw Coupling

6.2.1 Concept

In 3D walking, side-to-side tipping (i.e. excessive leaning) is a major stability hazard. One of the most effective control opportunities is lateral foot placement - placing the next swing leg sideways to avert a fall in that direction [96, 46]. The key insight in this chapter is that such lateral foot placement can be obtained by steering, i.e. yaw. Rotation around the vertical axis by itself has no effect, but in combination with forward progression, it suddenly affects the lateral foot placement. In other words, if during a forward walking motion one steers to the left, then automatically the next footfall will be placed more to the left side. Thus, yaw can be used for side-to-side stability, as long as there is a forward progression.

In this section, we will investigate two known examples of such advantageous lean-to-yaw coupling; a skateboard and a bicycle. These examples represent two different forms of lean-to-yaw coupling. The skateboard has a *kinematic* coupling; yaw and lean are combined in a single degree of freedom as explained in Section 6.2.2. The bicycle has a *dynamic* coupling; besides the sideways lean degree of freedom of the bicycle, there is a separate degree of freedom for the steering front fork. The bicycle study is presented in Section 6.2.3. Note that the remainder of the chapter addresses a *kinematic* lean-to-yaw coupling for walking machines, for which the skateboard is the most relevant analogy.

6.2.2 Skateboard

In the skateboard, the two sets of wheels are attached to the board via tilted steering axes (Fig. 6.1). Although the main purpose of this construction is to give the rider the ability to steer, here we'll show how this also provides stability. The following analysis is a simplified version of the skateboard stability analysis by Hubbard [79] in 1979.

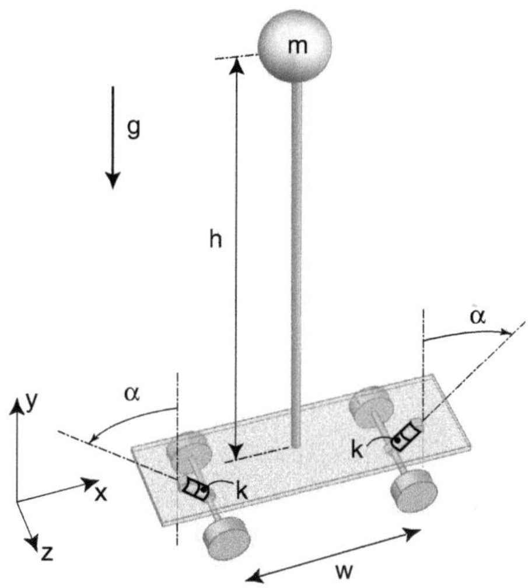

Fig. 6.1. Parameters of the skateboard model

The board and wheels are assumed to be massless and the height between the board and the floor is neglected. Also, we assume that there is always contact between all four wheels and the floor. The rider is modeled as a single point mass at height h above the floor, rigidly attached to the skateboard. The distance between the front and rear wheels is w. The steering axes are mounted at an angle α with respect to vertical such that sideways leaning of the rider results in steering in that direction. The steering axes are equipped with rotational springs with stiffness k. The model has fore-aft and sideways symmetry.

The skateboard is a non-holonomic system, i.e. it cannot slip sideways but it can move to a sideways position by a sequence of steering actions. Therefore it has a smaller velocity space (lean and ride) than coordinate space (lean, x- and y-position and orientation in plane). Here we will consider the linearized equations of motion in which the forward velocity can be considered as a parameter. The linearized model only has one degree of freedom, namely the sideways lean angle of the rider θ (Fig. 6.2). With zero forward velocity, the behavior equals that of an inverted pendulum with a de-stabilizing gravity torque and a stabilizing spring torque. Although the two springs act on the tilted joints, in their projected torques the tilt angle α cancels out.

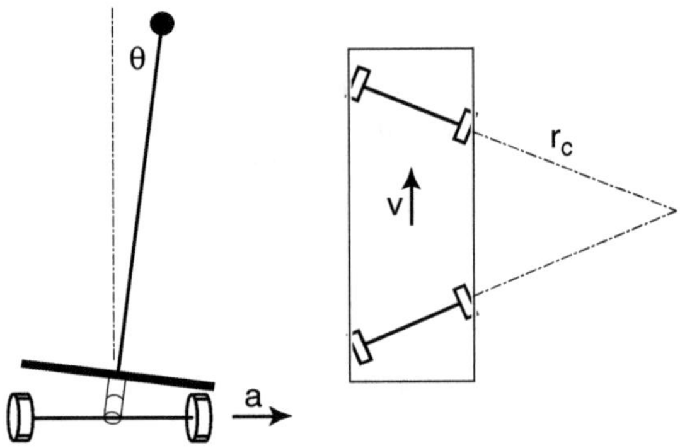

Fig. 6.2. Variables for the linearized skateboard model. **Left:** rear view. **Right:** top view.

When riding with a velocity v, the skateboard makes a turn if there is a non-zero lean θ. The velocity together with the radius of curvature r_c (Fig. 6.2) determine the sideways acceleration of the board according to

$$a = \frac{v^2}{r_c} = \frac{2v^2}{w \tan \alpha}\theta \tag{6.1}$$

Therefore, the total system equation describing the linearized model becomes

$$mh^2\ddot{\theta} + (2k - mgh + \frac{2mh}{w \tan \alpha}v^2)\theta = 0 \tag{6.2}$$

This result can be interpreted as follows. If the spring stiffness k is high enough to counteract the inverted pendulum instability, then the system is never unstable. If not, then it can always be made stable by its velocity. The larger angle α, the higher the required velocity. The critical velocity is:

$$v_{min} = \sqrt{\frac{(mgh - 2k)w \tan \alpha}{2mh}} \tag{6.3}$$

Note that the system can at best only be marginally stable. The introduction of damping in the steering axis could make it asymptotically stable. Also note that Eq. (6.3) suggests that it is best to make angle α equal to zero. In real life there is a lower limit to α dependent on the width of the skateboard because of the unilateral contact between the wheels and the floor.

6.2.3 Bicycle

Everybody knows that a bicycle is highly unstable at rest but can easily be stabilized at a moderate speed. Moreover, some uncontrolled bicycles can be asymptotically stable in a certain speed range. To demonstrate this phenomenon we will consider one of the simplest bicycle models: an uncontrolled bicycle with

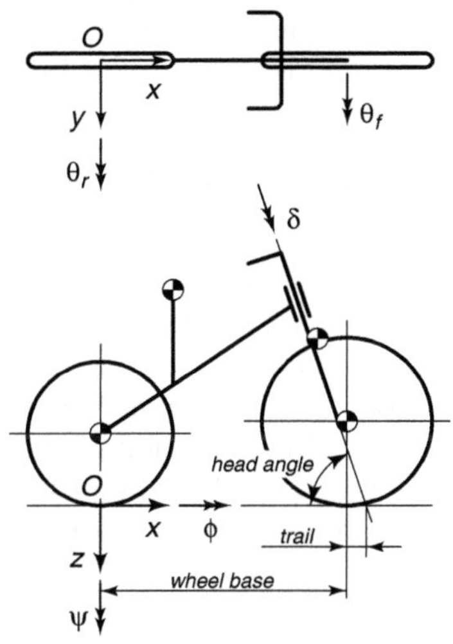

Fig. 6.3. Bicycle model together with the coordinate system, the degrees of freedom, and the parameters, from [140]

a rigid rider attached. This is an example of a dynamically coupled lean-to-yaw motion due to the hands-free operation of the bicycle. The following analysis is based on a recent bicycle benchmark publication by Schwab, Meijaard, and Papadopoulos [140].

The mechanical model of the bicycle consists of four rigid bodies, viz. the rear frame with the rider rigidly attached to it, the front frame being the front fork and handle bar assembly and the two knife-edge wheels. These bodies are interconnected by revolute hinges at the steering head between the rear frame and the front frame and at the two wheel hubs. The contact between the stiff non-slipping wheels and the flat level surface is modeled by holonomic constraints in the normal direction and by non-holonomic constraints in the longitudinal and lateral direction. There is no friction, apart from the idealized friction between the non-slipping wheels and the surface, nor propulsion and no rider control, the so-called hands free coasting operation.

The mechanical model of the bicycle has three degrees of freedom: the lean angle ϕ of the rear frame, the steering angle δ, and the rotation θ_r of the rear wheel with respect to the rear frame. The forward speed is $v = -\dot{\theta}_r R_{rw}$, where R_{rw} is the radius of the rear wheel. Due to the non-holonomic constraints there are four extra kinematic coordinates which describe, together with the degrees of freedom, the configuration of the system [139]. The four kinematic coordinates are taken here as the Cartesian coordinates x and y of the rear-wheel contact point, the yaw angle ψ of the rear frame, and the rotation θ_f of the front wheel with respect to the front frame. The dimensions and mechanical properties of the benchmark model are those of a regular 18 kg bicycle with an average 76 kg rider. For the complete set of parameters we refer to [140].

In this study, we consider the linearized model of the bike at constant forward speed. In the linearized model only two degrees of freedom remain, $\mathbf{q} = (\phi, \delta)^T$. The equations of motion are

$$\mathbf{M}\ddot{\mathbf{q}} + [\mathbf{C1} \cdot v]\dot{\mathbf{q}} + [\mathbf{K0} + \mathbf{K2} \cdot v^2]\mathbf{q} = \mathbf{f}\,, \tag{6.4}$$

with a constant mass matrix, $\mathbf{M}$, a "damping" matrix, $\mathbf{C1}$, which is proportional to the forward speed v, and a stiffness matrix which has a constant part, $\mathbf{K0}$, and a part, $\mathbf{K2}$, which is proportional to the square of the forward speed. Unfortunately the entries in the matrices are too complex and lengthy to express in a concise symbolic form. The benchmark paper [140] does present these entries in an algorithmic manner. Simplification of these expressions by neglecting so-called minor terms, as has been done by others in the past, would lead to incorrect results. Therefore we choose here to present typical values for the entries in the matrices, namely

$$\begin{aligned}\mathbf{M} &= \begin{bmatrix} 80.812 & 2.3234 \\ 2.3234 & 0.30127 \end{bmatrix}, \mathbf{C1} = \begin{bmatrix} 0 & 33.774 \\ -0.84823 & 1.7070 \end{bmatrix}, \\ \mathbf{K0} &= \begin{bmatrix} -794.12, & -25.739 \\ -25.739 & -8.1394 \end{bmatrix}, \mathbf{K2} = \begin{bmatrix} 0 & 76.406 \\ 0 & 2.6756 \end{bmatrix},\end{aligned} \tag{6.5}$$

where we use the standard units kg, m, and s. The forces $\mathbf{f}$ on the right-hand side are the action-reaction lean moment between the fixed space and the rear frame, and the action-reaction steering moment between the rear frame and the front frame. The latter is the torque that would be applied by a rider's hands or a controller. In the present study of an ordinary uncontrolled bicycle, both of these moments are taken to be zero.

To investigate the stability of the upright steady motion we start from the homogeneous linearized equations of motion (6.4). Next we assume for the small variations in the degrees of freedom an exponential motion with respect to time which then takes the form $\mathbf{q} = \mathbf{q}_0 \exp(\lambda t)$. This leads to an eigenvalue problem for which in this case the characteristic equation is a polynomial in the eigenvalues λ of order four. The coefficients in this polynomial are themselves polynomials in the forward speed v, since some coefficients of the linearized equations of motion have a linear or quadratic dependency on v. The solutions of the characteristic polynomial for a range of forward speeds are the root loci of the eigenvalues λ, which are shown in Figure 6.4. Eigenvalues with a positive real part correspond to unstable motions whereas eigenvalues with a negative real part result in asymptotically stable motions. Complex conjugated eigenvalues give rise to oscillatory motions.

For the bicycle model there are two significant eigenmodes, called capsize mode and weave mode. The capsize motion is a non-oscillatory motion in which, when unstable, the bicycle just falls over like a capsizing ship. The weave motion is an oscillatory motion in which the bicycle sways about the headed direction. Both eigenmodes show the dynamically coupled lean-to-yaw motion. At very low speed, $0 < v < 0.7$ m/s, there are two positive and two negative eigenvalues

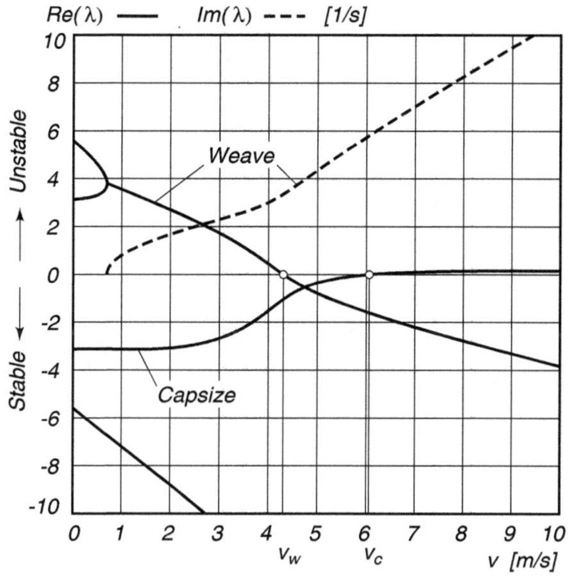

Fig. 6.4. Eigenvalues λ from the linearized stability analysis for the benchmark bicycle from Figure 6.3 and [140] where the solid lines correspond to the real part of the eigenvalues and the dashed line corresponds to the imaginary part of the eigenvalues, in the forward speed range of $0 \leq v \leq 10$ m/s. The zero crossings of the real part of the eigenvalues are for the weave motion at $v_w = 4.3$ m/s and for the capsize motion at $v_c = 6.1$ m/s, giving the bicycle an asymptotically stable speed range of $v_w < v < v_c$.

which correspond to an inverted pendulum-like motion of the bicycle. At $v = 0.7$ m/s two real eigenvalues become identical and start forming a conjugated pair; this is where the oscillatory weave motion emerges. At first this motion is unstable but above $v_w = 4.3$ m/s the weave motion becomes stable. After this bifurcation the frequency of the weave motion is almost proportional to the forward speed. Meanwhile the capsize motion, which was stable for low speed, becomes mildly unstable at $v_c = 6.1$ m/s. With further increase in speed, the capsize eigenvalue approaches zero.

We conclude that the bicycle model shows a dynamic lean-to-yaw coupling resulting in an asymptotically stable motion. Similar to the skateboard, with its kinematic coupling, a forward velocity above some critical value is required for this stabilizing effect.

6.3 Simplest Passive Walking Model with Lean-to-Yaw Coupling

6.3.1 Model

The purpose of the simulation model is to show the essential dynamic effects of kinematic lean-to-yaw coupling in walking systems. The simplest model for this purpose is a 3D cousin of Garcia *et al.*'s two-dimensional 'Simplest walking model' [54] which consisted of one finite point mass at the hip joint, two infinitesimally small point masses at the feet, and massless rigid links in between, interconnected with a frictionless hinge at the hip. Our model (Fig. 6.5) is a 3D extension of this; the hip has gained a finite width and the hip mass is divided into two point masses at the extremes of the massless hip axle. For numerical reasons the point masses at the feet of our model are not infinitesimally small but just very small. The degrees of freedom are the coordinates and the yaw and lean angles of the center of the hip axle $\{x_h, y_h, z_h, u_1, u_2\}$, the two leg pitch angles $\{u_3, u_4\}$, and the two ankle angles $\{u_5, u_6\}$. The ankle axes are mounted in the x-y plane at an angle α with respect to the vertical. Note that the ankle axes have no component in the z-direction, unlike conventional robot designs or the human ankle. The 'normal' ankle functionality, rotation around the z-axis, is realized by means of the roll-off motion of the feet.

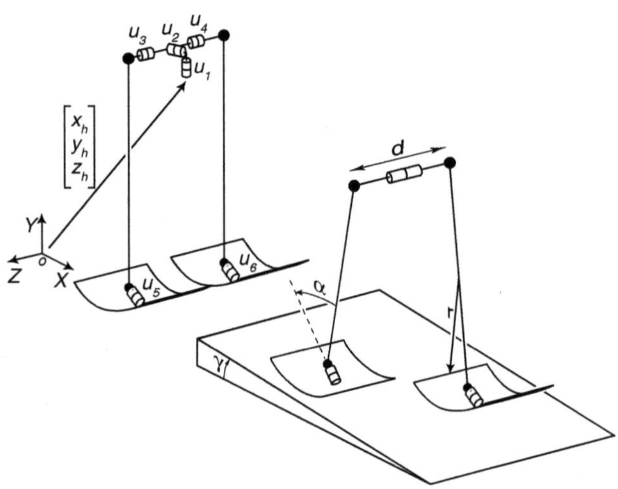

Fig. 6.5. Our simple walking model with ankle joints that couple lean to yaw. **Left:** degrees of freedom. **Right:** parameters.

The feet are (partial) cylinder shells with the cylinder axis perpendicular to the ankle axis. They are mounted such that the cylinder axis is in the leg (not displaced forward or backward) and that the total leg length (when standing upright) is independent of the cylinder radius. The foot contact is modeled as a perfectly rigid cylinder-plane contact with only one degree of freedom; pitch in a direction perpendicular to the cylinder axis. The width of the feet is not specified and is assumed to be sufficient to prevent sideways tipping over the edge. The feet in the model have a finite size but no inertia, so the foot of the swing leg contributes to the system dynamics only as a point mass. For the detection of heel contact, the swing foot is assumed to preserve its parallel orientation with respect to the floor surface. Heel contact itself is modeled as a rigid plastic impact with immediate and full contact to the floor, while at the same instant the previous stance foot loses contact.

The model walks down a shallow slope γ. All parameters are scaled so that gravity g, the leg length l and the total robot mass ($2m_{hip}$) are all equal to one[1]. The simulation results can be scaled back to obtain results for example for an earthly gravity regime. This scaling exposes the minimal set of adjustable parameters; hip width d, foot radius r, ankle mounting angle α, and slope angle γ as listed in Table 6.1. Additionally, a (small) torque can be exerted at the hip joint to move the swing leg quickly to a forward position, a feature that will prove necessary for stable walking (see Chapter 4 and Section 6.4).

Table 6.1. Parameters and their default values

hip width	d	0.3
foot radius	r	0.5
ankle mounting angle	α	0.55
slope angle	γ	0.01

6.3.2 Equations of Motion

The equations of motion for the system without foot contact constraints were generated with the method of virtual power as described in Chapter 2, where the twelve dependent degrees of freedom ($\{x, y, z\}$ for each point mass) were expressed in terms of the nine generalized coordinates $\{x_h, y_h, z_h, u_1..u_6\}$. In addition, there are three coordinates that change value only once per step; the foot roll-off direction ϕ and the foot contact location $\{x_c, z_c\}$. The five foot contact constraints are expressed as follows:

1. No yaw; the foot cylinder axis must remain perpendicular to its initial heading,
2. No lean; the foot cylinder axis must remain perpendicular to the normal of the floor,
3. Contact with plane; the center of the foot cylinder axis must remain a distance r (foot radius) above the floor,
4. No lateral slip
5. No forward slip; the forward disposition should match with the pitch angle.

These constraint conditions (either for the right foot or for the left foot depending on which one is in stance phase) were added to the equations of motion to obtain a system of Differential Algebraic Equations which solves for the 9 generalized coordinates and 5 Lagrange multipliers (one unknown contact force or torque per constraint condition). The impact equations for the heel strike event are

[1] Due to the applied scaling, most quantities in this text are dimensionless, which explains the frequent use of seemingly incomplete statements such as 'a velocity of 0.36'.

derived in the same systematic manner (see Chapter 2), in which the five foot contact constraints serve as contact conditions where impulsive forces can occur.

6.3.3 Simulation Procedure

The simulation procedure is a succession of simulations of walking steps which begin and end at the instant immediately after heel strike. Within one step, the system of DAE's is numerically integrated until heel strike is detected, followed by an impact calculation. The end state of the walker is then used as the starting state for a second step. After the second heel strike, the end state ($\{\mathbf{q}, \dot{\mathbf{q}}\}_{n+1}$) is compared to the initial state of the walker ($\{\mathbf{q}, \dot{\mathbf{q}}\}_n$) and the entire two steps can be summarized as the non-linear stride function $\mathbf{S}$ which maps the end states on the initial states:

$$\begin{bmatrix} \mathbf{q}_{n+1} \\ \dot{\mathbf{q}}_{n+1} \end{bmatrix} = \mathbf{S}\left(\begin{bmatrix} \mathbf{q}_n \\ \dot{\mathbf{q}}_n \end{bmatrix}\right) \tag{6.6}$$

Note that we do not apply the common procedure of state-mirroring at the end of the step, which is used in most of the 2D analyses to confine the simulation to only one walking step instead of two consecutive steps.

According to the Poincaré Mapping method, if the end state equals the initial state, we have found a *fixed point* representing a cyclic walking motion. The stability is determined by the effect of deviations $\boldsymbol{\epsilon}_n$ in the initial state on deviations $\boldsymbol{\epsilon}_{n+1}$ in the end state. For small deviations, we assume linearity around the fixed point, such that:

$$\boldsymbol{\epsilon}_{n+1} = \mathbf{J}\boldsymbol{\epsilon}_n \qquad \textit{with} \qquad \mathbf{J} = \frac{\partial \mathbf{S}}{\partial(\mathbf{q}_n, \dot{\mathbf{q}}_n)} \tag{6.7}$$

$\mathbf{J}$ is the Jacobian of the stride function $\mathbf{S}$ and is determined by performing the simulation procedure once for all deviations $\boldsymbol{\epsilon}_n$, one for each independent initial condition. The stability characteristics are described by the eigenvalues $\boldsymbol{\lambda}$ of the Jacobian $\mathbf{J}$; if all are smaller than 1 in magnitude, errors decay over subsequent strides. The smaller the eigenvalues, the faster the walker converges toward the fixed point. Note that the definition of eigenvalues here differs from Section 2; the analysis of the walking model is discrete ($-1 < \lambda < 1$ is stable) whereas the skateboard and bicycle analysis is continuous ($\lambda < 0$ is stable).

For the stability analysis, there are 6 relevant independent initial conditions $\{\mathbf{q}, \dot{\mathbf{q}}\}_n$ for the start of a stride. The 9 independent coordinates for the free model are reduced by 5 foot contact constraints to 4 independent degrees of freedom when one foot is in contact with the floor. The angle of the swing foot is not connected to any inertia, so this leaves only 3 independent degrees of freedom and thus 6 states, namely $\{u_3, u_4, u_6, \dot{u}_3, \dot{u}_4, \dot{u}_6\}$ for left stance. The Poincaré Section removes one state (one of the hip angles) so that there would be 5 independent initial conditions. In addition, there are three coordinates that change only once per step, namely x_c, z_c and ϕ. The linear coordinates x_c and z_c are not relevant to the walking motion, but the foot roll-off direction ϕ is, so it must be added to

Table 6.2. Fixed point initial conditions for left stance, valid for the parameter values in Table 6.1

foot heading	ϕ_2	0.0016
stance leg angle	u_4	0.155
stance ankle angle	u_6	−0.0041
stance leg angular velocity	$\dot{u}_4$	−0.42
stance ankle angular velocity	$\dot{u}_6$	−0.070
swing leg angular velocity	$\dot{u}_3$	−0.42

the set of independent initial conditions. Therefore, in total there are 6 relevant independent initial conditions as listed in Table 6.2.

6.4 Simulation Results

This section presents the behavior and stability of five simulation models of increasing complexity. The first model (Fig. 6.6A) is the two-dimensional 'simplest walking model' of Garcia *et al.* [54]. This fully passive model has straight legs with point feet (with infinitesimally small point masses) and a large point mass at the hip. The second model (Fig. 6.6B) is similar except that it has arc feet instead of point feet. The third model (Fig. 6.6C) is equal to the second model except that it is three-dimensional. It has no hip width but it can move out-of-plane. Also, this model is equipped with the proposed ankle joints. The fourth model (Fig. 6.6D) differs in that it has a non-zero hip width, and the fifth model (Fig. 6.6D) has an additional actuator at the hip joint (Chapter 4).

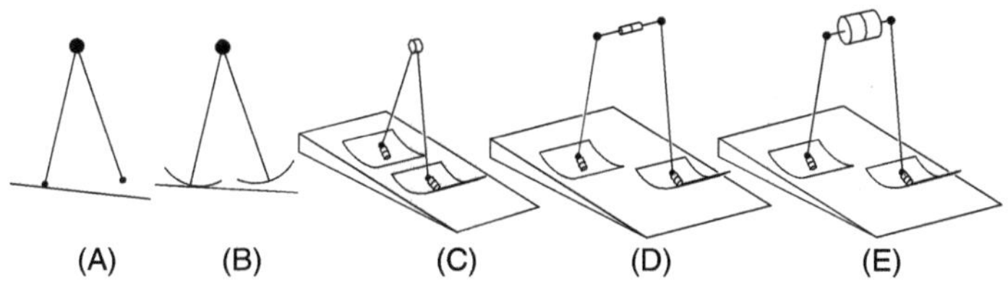

Fig. 6.6. Models with increasing complexity. A) 2D point foot walker [54], B) 2D arc foot walker, C) flat 3D walker (passive or active), D) passive 3D walker with finite hip width, E) active 3D walker.

6.4.1 Fully Passive Model

Garcia *et al.* [54] researched the simplest walking model in 2D that could still demonstrate a passive walking motion, Fig. 6.6A. The model consists of three point masses, one of mass 1 at the hip and two infinitesimally small point masses at the feet, with rigid, massless links as legs interconnected with a frictionless hinge. Therefore it has two degrees of freedom when in stance phase and thus

$2 \times 2 - 1 = 3$ independent initial conditions for a step starting with the rear leg just leaving the floor. With its point feet (no radius of curvature), the simplest walking model is a special 2D case of the model presented in this chapter.

The model shows a stable walking pattern when walking on a shallow slope with a downward angle smaller than 0.015 (rad) [54]. For example, Table 6.3 presents the eigenvalues $\boldsymbol{\lambda}$ for the cyclic motion that exists for a typical slope of 0.004 (rad). It has been shown, however, that the simplest walking model is highly susceptible to disturbances (see Chapter 4), and that this sensitivity can be greatly reduced by the application of arc feet with a substantial radius [185]. Therefore we continue this chapter with a model with arc feet with a radius of (a somewhat arbitrarily chosen) 0.5 times the leg length. For comparison within this chapter and with previous publications [141, 184], the arc foot model is given the same step length and approximately the same velocity as the point foot model by adjustment of the slope angle. The step length is determined by the initial stance leg angle θ_0. On a slope of $\gamma = 0.004$ (rad) the point foot model has a limit cycle with $\theta_0 = 0.15$ (rad). The arc foot walker only needs a slope of $\gamma = 0.00058$ (rad) for the same step length, i.e. it is about 7 times more efficient while walking at the approximate same velocity, see Fig. 6.7. The eigenvalues have moved closer to 1 and thus do not suggest a stability improvement (Table 6.3). However, we performed a crude analysis of the basin of attraction which showed that the arc foot walker can handle deviations from the initial conditions of 8% versus 2% for the point foot walker, indicating a better practical applicability of the model.

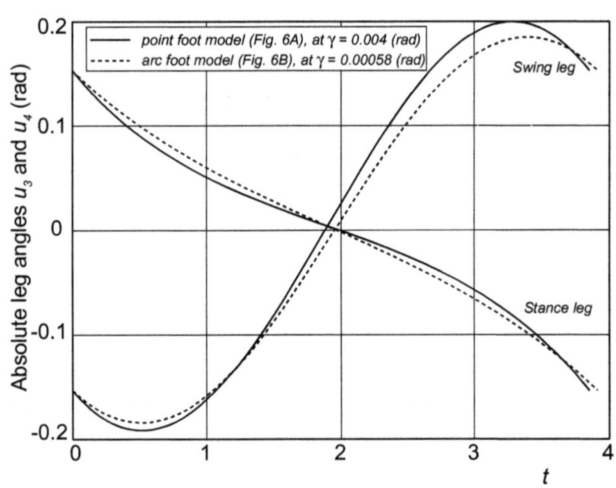

Fig. 6.7. Leg angles versus time for a point foot model (Fig. 6.6A) and an arc foot model (Fig. 6.6B) with a foot radius $r = 0.5$. The displayed cyclic walking motion (only one step is shown) is valid both for 2D models and for 3D models with zero hip width (Fig. 6.6C).

Table 6.3. Eigenvalues of the cyclic walking motion for the simplest walking model with point feet, a 2D walker with arc feet, and a flat ($d = 0$) 3D walker with ankle axes oriented at $\alpha = 0.15$ (rad). Note that all eigenvalues result from *two* successive steps, i.e. one *stride*.

eigenvalues	2D point foot walker (Fig. 6.6A) at $\gamma = 0.004$	2D arc foot walker (Fig. 6.6B) at $\gamma = 0.00058$	3D with arc feet (Fig. 6.6C) at $\gamma = 0.00058$
λ_1	$-0.3 + 0.27i$	0.69	0.69
λ_2	$-0.3 - 0.27i$	0.12	0.12
λ_3	0	0	0
λ_4	-	-	0.72
λ_5	-	-	0.51
λ_6	-	-	0

An interpretation of the eigenvectors corresponding to the last set of eigenvalues in Table 6.3 has shown that the first three eigenvalues are indeed only related to the fore-aft motions whereas the last three eigenvalues are related to 3D motions in which both the sideways and fore-aft direction are present. The most important conclusion to be drawn from Table 6.3 is that there exists a stable 3D walking motion for our model with no hip width and an almost upright ankle axis ($\alpha = 0.15$ rad).

Further research shows that the eigenvalues are highly sensitive to the orientation of the ankle axis and that only a small region leads to stable motions, see Fig. 6.8. The first three eigenvalues are not a function of the ankle axis orientation as it only influences 3D motions. The small region of stable values for α is located around the maximal stance leg angle (0.15 rad, see Fig. 6.7). This means that, during the walking motion, the orientation of the ankle axis will move from 0.3 to 0 (rad) with respect to an absolute reference frame, i.e. it will have a completely vertical orientation at the end of each step. Obviously, in this orientation the ankle joint is no longer a "lean-to-yaw" coupling, but it has become a pure yaw degree of freedom. This eliminates the stabilizing effect for two reasons. First, there is no coupling so the yaw motion is not related to the rest of the walking motion, resulting in uncorrelated (and thus de-stabilizing) rotations around the vertical axis. Second, more severely, the loss of the lean degree of freedom in the ankle joint will most likely result in sideways tipping over on the inside edge of the foot, unless very wide feet are applied. Recall that a similar lower boundary exists for a realistic skateboard model.

We hypothesized that a finite hip width $d > 0$ might add to stability. A graph of the eigenvalues as a function of both the ankle orientation α and the hip width d is sketched in Fig. 6.9. The sketch is based on a number of cross-sections in the parameter space of α and d. The stable region for α narrows down with an increasing d up to $d \approx 0.05$, beyond which no stable solutions were found for any

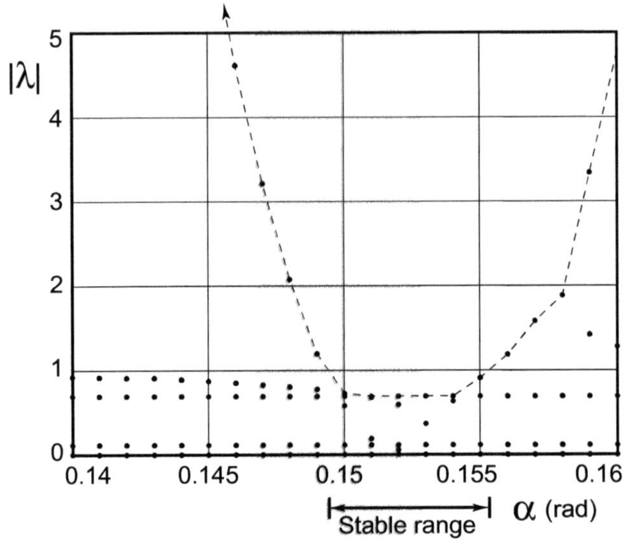

Fig. 6.8. Typical plot of the absolute values of the eigenvalues $|\lambda|$ as a function of the ankle joint orientation α for a fully passive 3D model (Fig. 6.6C). This plot is generated with hip width $d = 0$ and foot radius $r = 0.5$, walking on a slope of $\gamma = 0.00058$ (rad).

α. A search up to $d = 0.2$ provided ever increasing eigenvalues, so we extrapolate the result to conclude that no stable 3D walking motions exist for a hip width of more than 0.05 times the leg length.

Another hypothesis was that a steeper slope might improve the stability; on a steeper slope the passive walker will take larger steps and thus walk faster, and the skateboard and bike models predict a beneficial stability effect for higher velocities. However, the simulation has shown only marginal effects. Up to the maximal slope of 0.046 (rad) beyond which no stable motions exist, the 3D graph in Fig. 6.9 remains similar in shape. The stable region shifts to higher values of α along with the increase in step length. For example, a slope increase from 0.00058 to 0.004 (rad) causes the initial stance leg angle to increase from 0.15 to 0.29 (rad) and also shifts the stable values for α from around 0.15 to around 0.29 (rad).

These marginal stability result, together with the required vertical orientation of the ankle axis, indicate that the fully passive model is not sufficiently applicable for real-world prototypes and warrant a search for a model with stable behavior for larger values of α and d.

6.4.2 Model with Hip Actuation

In this subsection we propose to improve the overall (3D) walking behavior through the addition of a stabilizing feature that was originally intended for 2D machines. In two dimensions, the most persistent failure is a fall forward

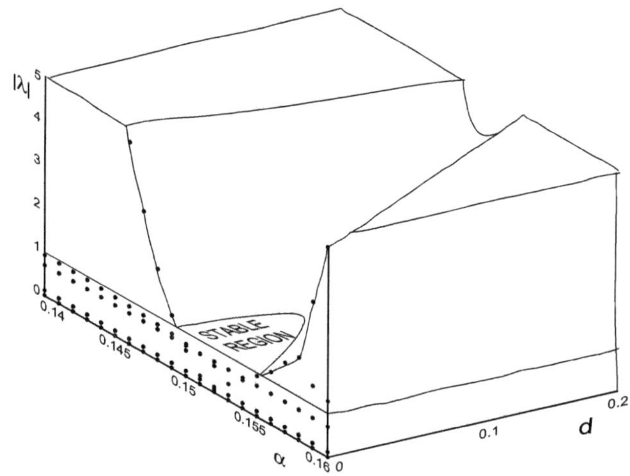

Fig. 6.9. A sketch of the dependency of the eigenvalues as a function of ankle joint orientation α and hip width d for a model (Fig. 6.6D) with foot radius $r = 0.5$ and slope angle $\gamma = 0.00058$ (rad). The left face of the figure is equal to Fig. 6.8.

which can be averted by simply accelerating the swing leg to bring it quickly to a forward position and subsequently keeping it there (Chapter 4). This simple form of swing leg control can be implemented on our 3D model without energy implications; the swing leg is nearly massless, so any control action can be applied (almost) without reaction torques to the rest of the model. Therefore, the model still requires a downhill slope for a sustained walking motion. This also means that the swing leg can be moved arbitrarily fast; in the remainder of this chapter the swing leg is assumed (and simulated) always to be in the forward position before heel strike occurs.

We can arbitrarily set the forward angle that the swing leg is quickly moved to. Let's maintain the value of 0.15 (rad) as initial stance leg angle in accordance with the passive models described earlier. The controller therefore has to move the swing leg quickly to an inter-leg angle of $2 \cdot 0.15 = 0.3$ (rad). The immediate effect of the controller is that disturbances on the initial swing leg velocity do not affect the end state, thus one of the eigenvalues of the model becomes zero.

Looking at the behavior for $\gamma = 0.00058$, there seems to be little difference between the active model (Fig. 6.10A) and the passive model (Fig. 6.8). The reason is that the main improvements are to be found for larger slopes. For the active model, a change of slope does not lead to a change of step length, while the swing leg control ensures that the model cannot fall forward, so any arbitrarily steep slope can be used. These two effects of the swing leg control together result in a much more favorable behavior. In contrast to the passive model, for the active model an increase of γ does not lead to a *translation* of the stable region for α but rather to an *increase* of this region, shown in Fig. 6.10B.

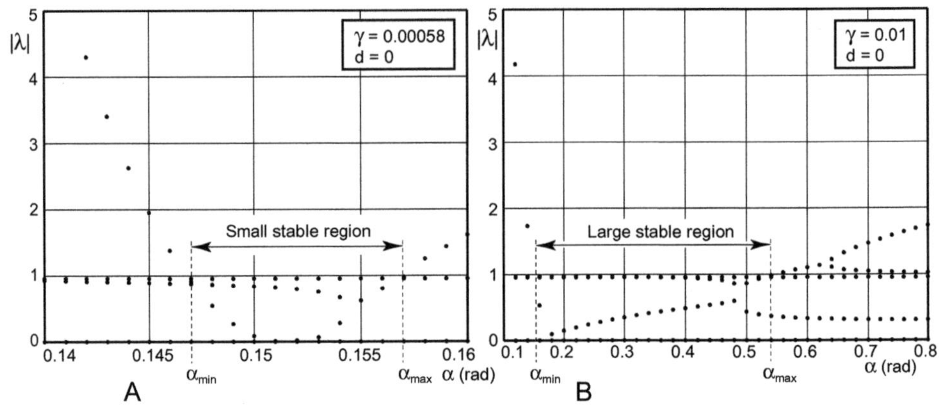

Fig. 6.10. Typical plot of the absolute values of the eigenvalues $|\lambda|$ as a function of the ankle joint orientation α for a 3D model with simple swing leg control (Fig. 6.6C). This plot is generated with hip width $d = 0$ and foot radius $r = 0.5$. **A:** walking on a slope of $\gamma = 0.00058$ (rad), **B:** walking on a slope of $\gamma = 0.01$ (rad). The two graphs exemplify that the upper boundary for α increases with an increasing γ; the steeper the slope, the larger the region of stability. Note the difference in scale of the two graphs.

The stable region in Fig. 6.10B has two independent boundaries. The lower boundary for α is directly related to the step length; if α is lower than 0.15, the ankle axis would reach a vertical orientation at the end of the step which leads to directional instability. With a fixed step length due to the swing leg controller, this lower boundary for α is more or less static. The upper boundary is directly related to the slope angle; the steeper the slope, the higher α can be. Or, in other words, for a given value of α the slope angle must be above a certain critical value for stable walking. In Fig. 6.11 this is depicted for $\alpha = 0.55$, the upper boundary from Fig. 6.10B. Note the correlation between the two graphs; the upper boundary $\alpha = 0.55$ in Fig. 6.10B corresponds to the stability boundary $\gamma = 0.01$ in Fig. 6.11.

With respect to the passive model, the active model has a much larger stable region; when walking on a slope of $\gamma = 0.01$ (rad), the ankle orientation angle can be anything between $0.15 < \alpha < 0.55$ (rad) for stable walking. This gives good hopes for models with a finite hip width. Fig. 6.12 presents the eigenvalues as a function of both α and the hip width d, where the left front plane equals Fig. 6.10B. Apparently, for the active model an increase of the hip width is even *beneficial* to the stability, in sharp contrast with the passive model.

6.4.3 Stability Versus Velocity

For the active model, it is easy to find a parameter combination that results in stable walking. Fig. 6.11 basically suggests that for any parameter combination, one can find stability by increasing the slope angle past a critical value. This

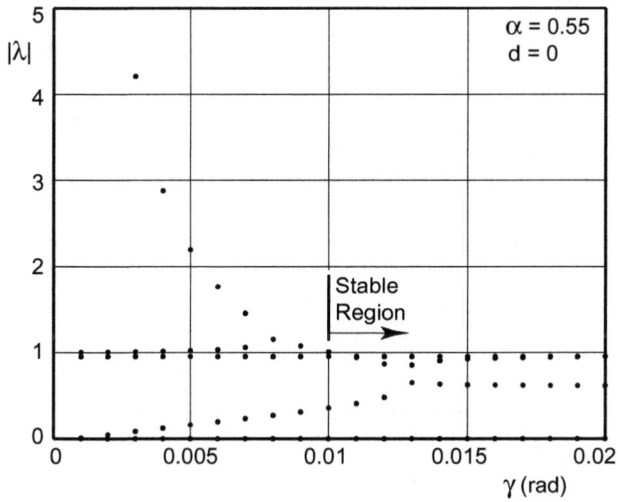

Fig. 6.11. Typical plot of the absolute values of the eigenvalues $|\lambda|$ as a function of slope angle γ for a 3D model with simple swing leg control (Fig. 6.6C). This plot is generated with hip width $d = 0$, a foot radius $r = 0.5$, and an ankle joint orientation $\alpha = 0.55$ (rad). For $\gamma > 0.01$ the model is stable. This boundary corresponds to the the upper stability boundary in the right graph of Fig. 6.10.

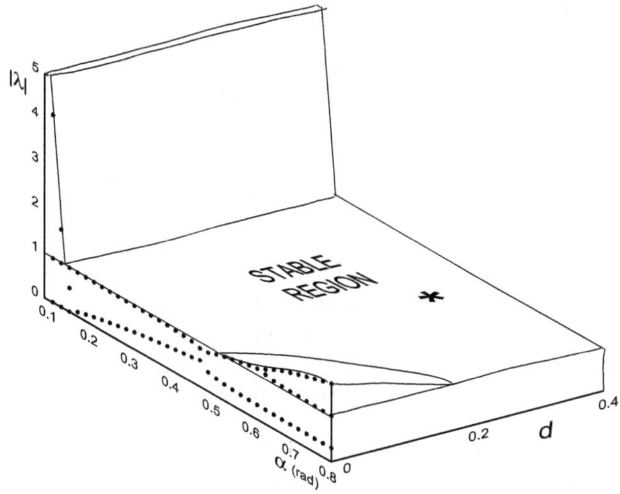

Fig. 6.12. Sketch of the eigenvalues of the active model as a function of ankle joint orientation α and hip width d for a model (Fig. 6.6E) with foot radius $r = 0.5$ and slope angle $\gamma = 0.01$ (rad). The left face is equal to the right graph of Fig. 6.10. The asterisk indicates the parameter set of Table 6.1.

effect has the same feel to it as the velocity relation in skateboards and bicycles, in which for most parameter values there exists a critical velocity above which stable motions occur. Therefore it is interesting to investigate the relation between the slope angle, the walker's velocity, and its stability, to answer the question: 'Is there a direct relationship between velocity and stability?'

To answer that question, we need to bring r into the equation because the velocity is determined by γ and r for a given step length. For manageability of the model and calculations, we will answer this question only for the *flat* version of the 3D model, i.e. $d = 0$ (Fig. 6.6C). The result is stunning; according to Fig. 6.13 there is almost a one-to-one relationship between the velocity and the stability, irrespective of the specific values of r and γ that cause that velocity!

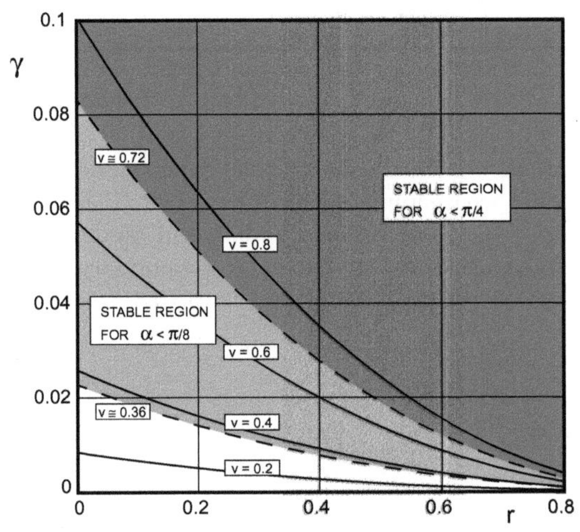

Fig. 6.13. Stability regions and walking velocity as a function of the slope angle γ and the foot radius r for an active, flat 3D model (Fig. 6.6C). The velocity is determined by the parameters r and γ and is independent of α. The stability is dependent on all three parameters. The graph shows that a more vertical ankle axis (smaller α) provides stability for lower velocities. Interestingly, the stability boundaries are almost completely coincident with lines of constant velocity; the (dashed) boundary of the (light grey) stability region for $\alpha < \pi/8$ is almost exactly equal to the line of constant velocity for $v = 0.36$ (not shown). Similarly, the (dashed) boundary of the (dark grey) stability region for $\alpha < \pi/4$ is almost exactly equal to $v = 0.72$ (not shown). Therefore we conclude that stability can be seen as a function of the velocity independent of the particular slope and foot radius that cause the velocity. Note that the velocity lines are obtained with the numerical simulation, not with the algebraic approximation in Eq. (6.8).

The result in Fig. 6.13 was obtained as follows. First, the plot contains contour lines of constant velocity. The walking velocity is a result of the gravitational energy input and the energy loss at the heel strike impact which are in balance when the walker is in a limit cycle. An analysis of the energy balance (not shown here for brevity) leads to the following approximate relationship:

$$\mathrm{v} \approx \frac{1}{(1-r)}\sqrt{\frac{\gamma}{\theta}} \tag{6.8}$$

where v is the walking velocity and θ is the stance leg angle at the start of the step (equal to half of the preset inter-leg angle). This approximation ignores the velocity decrease at midstance and therefore slightly overestimates the walking velocity. Eq. (6.8) clearly shows that if r equals the leg length 1, then the walker could have any velocity at a slope of 0 while no equilibrium exists for any other slope angle, because no energy is lost during heel strike. In Fig. 6.13 we used the exact velocities (from the non-linear simulations) rather than the approximation in Eq. 6.8.

The second ingredient of Fig. 6.13 is the shape of the region of stability. The figure shows for two different values of the ankle joint orientation α what combinations of values for the slope angle γ and the foot radius r lead to stable walking. The figure shows what was already known from Fig. 6.10; a steeper slope (larger γ) allows for a more horizontal ankle joint (larger α). The additional information in Fig. 6.13 is how the stability boundary depends on both γ and r, which apparently coincides with contour lines of constant velocity. A heuristic formula for the boundary value of α as a function of the velocity v can easily be extracted from the data scattered in this chapter. From Figures 6.10 and 6.13 we can extract four data points that show an almost one-to-one linear relationship, $v_{min} \approx \alpha$.

6.4.4 Walking and Steering

The previous subsections have shown that it is easy to find stable parameter combinations for the active model. This subsection will investigate the resultant walking motion for one characteristic set of parameter values (Table 6.1).

For a steady walk, the projection of the center of mass and the footprints are shown in Fig. 6.14. The step length of 0.3 times the leg length is a direct result of the swing leg controller. The center of mass makes sideways excursions of ± 0.008 times the leg length.

The model's inherent stability means that it will react to any (not too large) disturbance by asymptotically moving back to its steady limit cycle. This fact can be used for intentional steering; if the model were placed on the slope in a direction other than steepest descent, it will automatically steer toward that direction. Or, even more useful, a sideways mass offset will also induce steering in that direction. In Fig. 6.15, which contains a sequence of 500 walking steps, the center of mass was displaced slightly (0.006 times the leg length) to the right. The result is that the model asymptotically moves toward a heading direction of 0.64 (rad) with respect to the direction of steepest descent. With its new heading, the model has found balance between the mass offset to the right and the offset effect of the slope to the left.

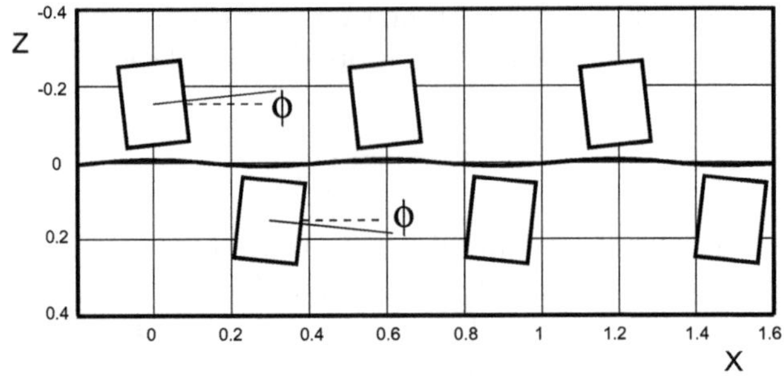

Fig. 6.14. Projection of the center of mass on the floor together with footprints for the full model (Fig. 6.6E). The grid on the floor is in units of leg length. The model walks from left to right in a steady motion. The footprint direction ϕ is exaggerated; the true value $\phi = 0.00155$ (see Table 6.2) would be invisible in the figure.

The model is robust enough to handle a much larger mass offset. Even if the center of mass is displaced sideways with 0.05 times the leg length, a steady (though somewhat limping) walking motion exists. With this offset, the model will turn with a radius of about 8 times the leg length, as shown with the dashed line in Fig. 6.15. It stops and falls not because of the direct effect of the mass offset, but because it has turned more than 90 degrees and thus receives no energy input. If the slope would turn with the walker, it would walk indefinitely in circles.

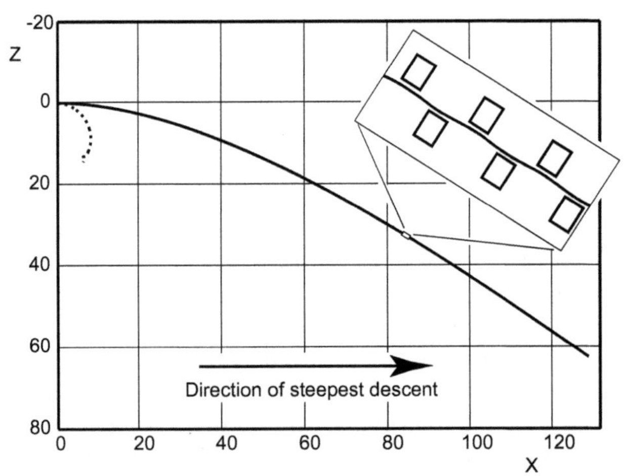

Fig. 6.15. Projection of the center of mass on the floor together with footprints for 500 steps. The model has a mass offset to the right and thus steers asymptotically toward a direction in which the sideways slope effect is in balance with the effect of the mass offset. A much larger mass offset results in a tighter turn as shown with the dashed line. After a turn of more than 90 degrees, the walker receives no energy input and eventually stops and falls.

As a final stability test we investigated the disturbance rejection of the model. The model was started with the initial conditions for the steady walking motion plus an error on one of them. The model is able to recover from an increase of at least 200% or a decrease of 100% on any of the initial conditions, except for the velocity of the stance leg angle. The stance leg's angular velocity can only be decreased with 50%, otherwise the model falls backward. All in all, the model predicts great potential for practically applicable prototypes.

6.5 Applicability in Real Walking Robots

The simulations predict successful walking for prototypes with the special ankle joint that couples falling sideways (lean) to turning in that direction (yaw). The solution to any instability is to increase the walking velocity. There is, however, a practical limit to the walking velocity. The simulation assumes that the (almost massless) swing leg is always in time to catch the walker for its next step, but a physical swing leg with substantial mass cannot move instantaneously. Its velocity is not only limited by practical considerations (actuator capacity), but also by the fact that its reaction torque might exceed the friction torques that the stance foot can supply.

Our model is not equipped with springs in the ankle joints for the sake of simplicity. The skateboard analysis, however, predicts a beneficial influence of such springs. For the development of a practical prototype, it is recommended to investigate the possible stability benefit that such springs can provide for walking robots.

The implication of this simulation study for the creation of stable prototypes goes beyond the concept of the tilted ankle joint. Even if proposed ankle joint is not implemented in a prototype, it is still possible to benefit from the idea behind it; any control algorithm anywhere in the body can have similar beneficial stability effects, as long as its effect is a lean-to-yaw coupling. We expect that such an effect is also present in the human body, albeit well masked by the simultaneous presence of two other control strategies for sideways stability, namely sideways foot placement [45, 46, 12] and inertial reaction torques from the upper body.

In the future, we hope to gain more insight in the similarity between a kinematic lean-to-yaw coupling as studied in this chapter, and a dynamic coupling as presented in the bicycle example. A promising lead is the stable walking behavior that was found in a 3D rimless wheel [31]. Another lead is provided by Kuo [96] who found that lateral stability can be obtained through fore-aft leg motions, which should also be considered as a dynamic lean-to-yaw coupling.

6.6 Prototype Design

The ankle joint is tested in the prototype 'Denise' (Fig. 6.16) [178], which weighs 8 kg (Table 6.4), measures h x b = 1.5 x 0.3 m (Fig. 6.17) and walks at 0.4 m/s (0.8 s per step). Denise is a direct successor of Max (Chapter 5) and has

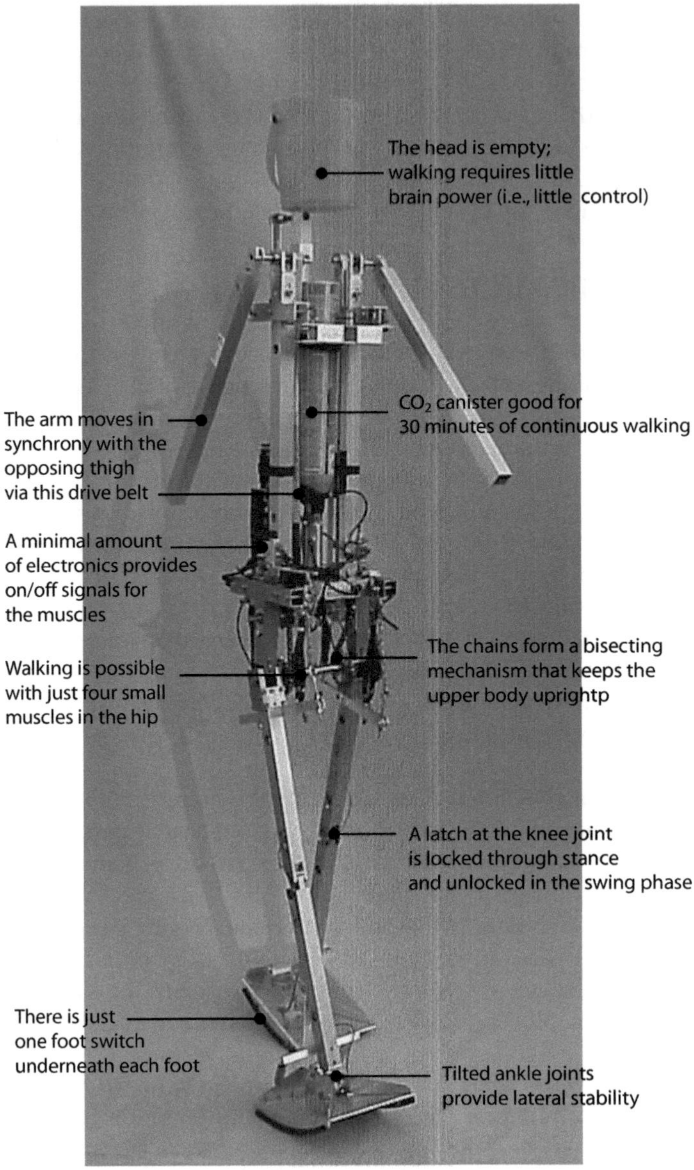

Fig. 6.16. Denise, minimalistic 3D dynamic walking robot with 5 degrees of freedom; two ankles, two knees, one at the hip, and the arms are rigidly coupled to the hip angle

the same controllable knee latches, hip actuation, and bisecting hip mechanism (Fig. 6.18). The muscle properties are given in Table 6.5. The hip joint has a maximal extension stop made of steel cables which limit the inter-leg angle to +/- 40°. This is an important feature for stable walking, see Chapter 4. The prototype has five internal degrees of freedom (Fig. 6.16); two at the ankles, two

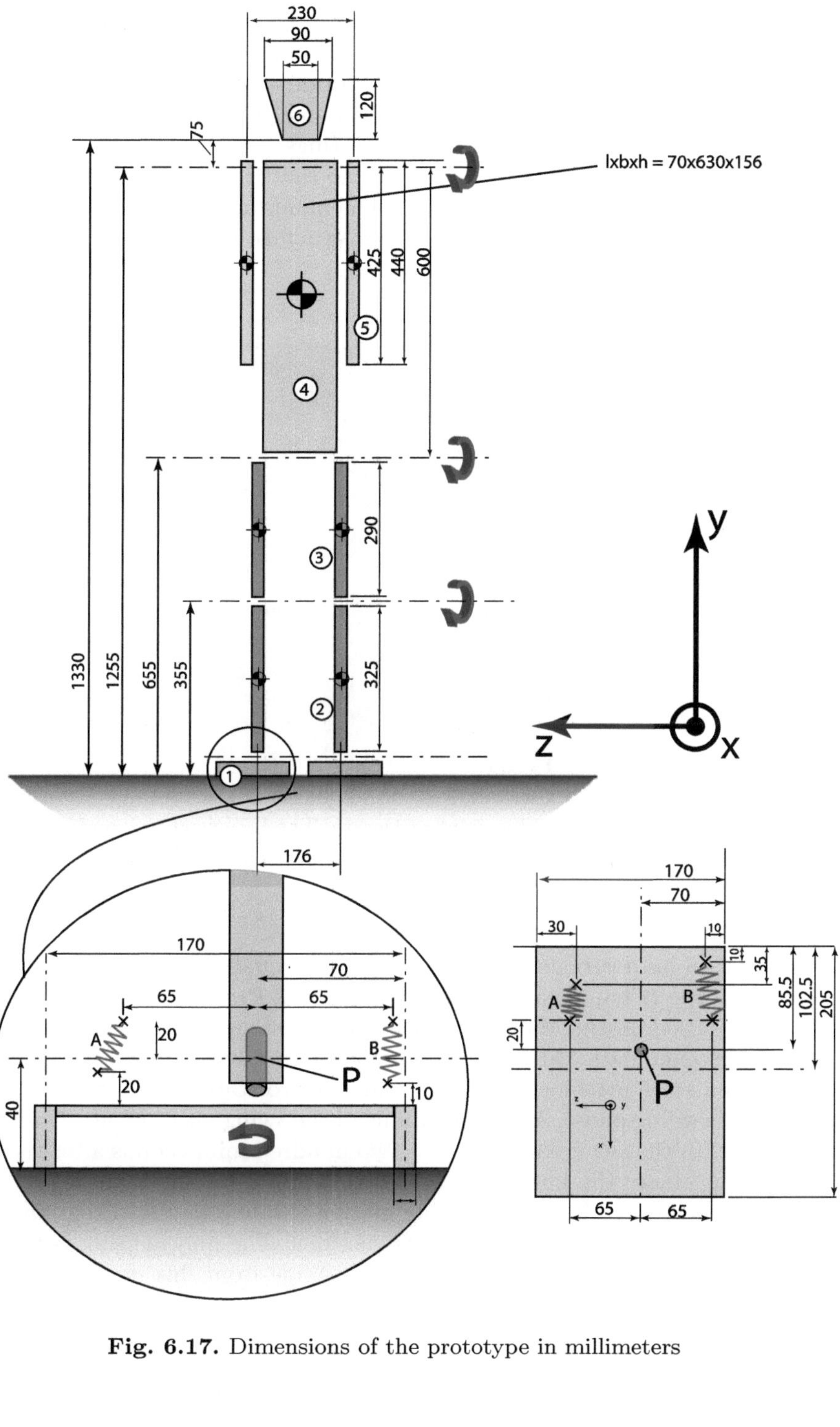

Fig. 6.17. Dimensions of the prototype in millimeters

at the knees, and one at the hip (the arms are mechanically connected to the opposing leg). The ankle joints are mounted in the non-human orientation as proposed above, namely pointing forward and downward without a component in the lateral direction (Fig. 6.19), making an angle $\alpha = 25°$ with the leg. There is a rotational stiffness by means of two springs, which helps keep the robot upright. The springs are as stiff as possible so that the foot just remains in full contact with the floor during the motion. The mounting positions for the springs are shown in Fig. 6.17. Their stiffness is 912 N/m for the outside spring and 5501 N/m for the inside spring.

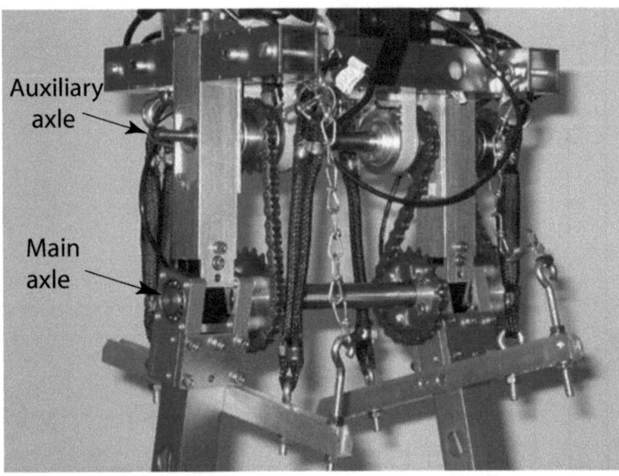

Fig. 6.18. Photograph of the hip joint of Denise. The bicycle chains and chainwheels together with the auxiliary axle form the bisecting mechanism. The picture also shows the McKibben muscles.

The control hardware and software is exactly equal to those used in the previous prototype (Chapter 5). We used a Universal Processor Board from Multi Motions® (based on the Microchip® PIC16F877 micro-controller) which opens or closes the pneumatic valves based on foot-contact switch signals. The control program is a state machine with two groups of two states: either the left or the right leg is in swing phase. At the beginning of the swing phase (State 2 and State 4 in Fig. 6.20), the swing knee is bent. Two hundred milliseconds after the start of the swing phase, the knee latch is closed, waiting for the lower leg to reach full extension through its passive swing motion (State 3 and 1). Programmed in assembly, this amounts to about 30 instructions. The only sensing is the time of foot contact, which occurs once per step. Aside from this input, there is no feedback to counter disturbances; the stability is inherent in the mechanism's motion.

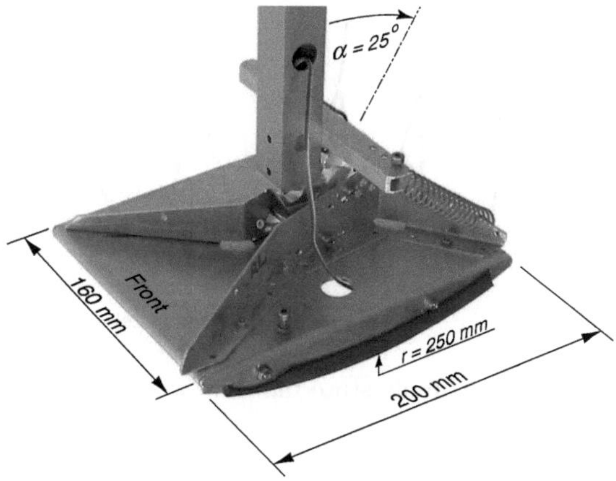

Fig. 6.19. Details of the (right) foot. The ankle joint is placed at an angle of 25° with respect to the lower leg.

Table 6.4. Mass distribution of Denise. The part numbers and x-y-z coordinates refer to Fig. 6.17. The center of mass coordinates are measured from a global reference frame with the origin at the floor between the feet. The inertia is measure with respect to each component's center of mass.

Part	mass (kg)	$\{x, y, z\}_{com}$ (mm)	$\{I_{xx}, I_{yy}, I_{zz}\}$ (kg·mm^2)
1 (Foot)	0.404	$\{5, 5, \pm 88\}$	$\{459, 454, 409\}$
2 (Shin)	0.215	$\{0, 221, \pm 88\}$	$\{2297, 2276, 35\}$
3 (Thigh)	0.295	$\{0, 509, \pm 88\}$	$\{2580, 2551, 48\}$
4 (Body)	5.4 - 7.0	$\{0, 957, 0\}$	$\{2.0, 1.9, 0.14\} \cdot 10^5$
5 (Arm)	0.229	$\{0, 1037, \pm 115\}$	$\{4269, 4247, 37\}$

Table 6.5. Properties of the McKibben muscles used in the hip joint of the prototype Denise. The values are obtained from fits of a full multibody simulation model to experimental measurement data.

Muscle length at zero force	131.5 mm
Stiffness at 6 atm	5 N/mm
Damping (fitted from motion data)	0.2 Ns/mm
Time constant for pressure increase	0.09 s
Time constant for pressure decrease	0.05 s

6.7 Prototype Experiments

6.7.1 Gait Characteristics

The main result to report is that Denise walks stably. An illustration of the walking motion is given in Fig. 6.21. For video evidence, please refer to our

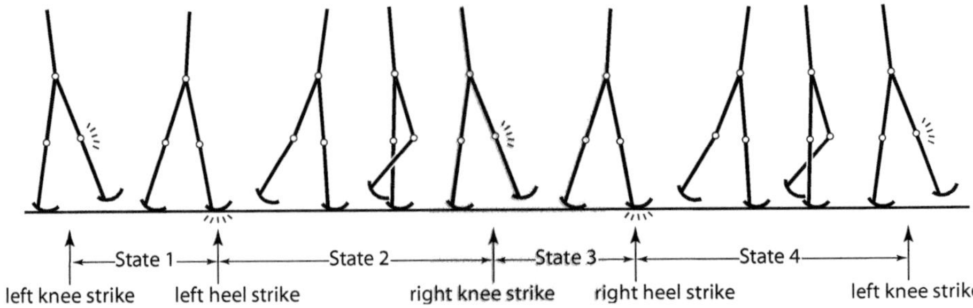

Fig. 6.20. The controller is a simple state machine which cycles through four states

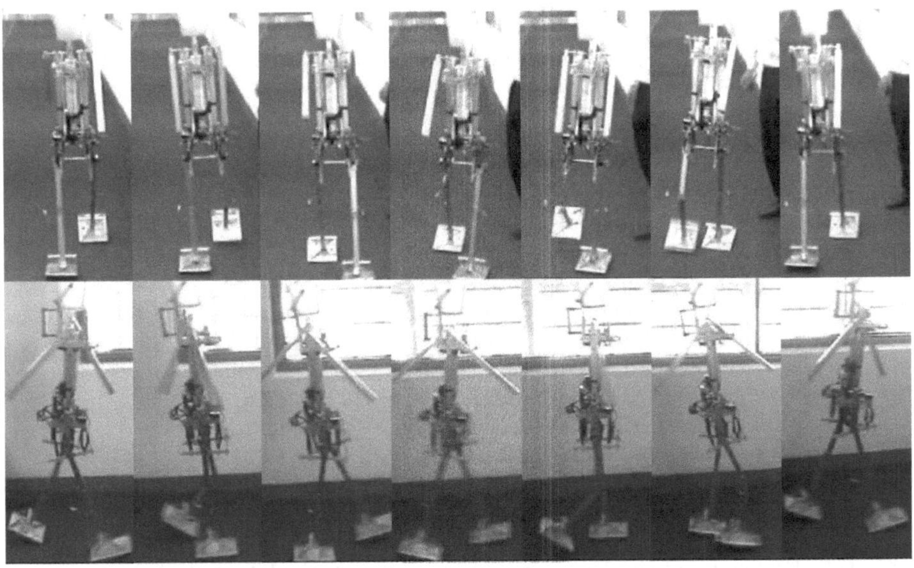

Fig. 6.21. Video stills of Denise walking, two steps (one stride). The video shows a slight turn to the right as a result of a disturbance in the floor.

website [1]. Unfortunately, we cannot provide a full experimental validation of the simulation results. The simulation results show that the stability can be improved by increasing the walking velocity and by placing ankle joint more vertical. As we were limited by the available pneumatic power due to flow limits, we needed both of these solutions for finding a successful walking motion. Thus, we can only present the inverse claim: the prototype does *not* walk stably 1) when using less than maximal pneumatic power, 2) when using a less vertical ankle joint (we tried ankle joints of 45° and 25° with respect to vertical, it failed with the first and succeeded with the latter). A redesign of the pneumatic system would be required before we can provide more quantitative support of the simulation results.

In the remainder of this chapter, we will report how the robot moves for comparison with future robots. We present not only the experimental results, but also the results of a complete multibody simulations performed in ADAMS. By matching the behavior of the simulation to the experimental results, we have more confidence that we understand how the prototype behaves. For example, the muscle properties of Table 6.5 were obtained by tuning the walking behavior of the ADAMS model to match the experiments. The experimental data was obtained with 3D motion capture using a VICON system with help from Brian Moyer at the University of Pittsburgh.

A high-level way of averaging the experimental results and comparing them with the simulation is by investigating the gait characteristics (step length and step frequency). The results are shown in Table 6.6. The experimental results are obtained from over 150 steps from several trials. Point-wise, the simulation results fall well within the standard deviation of the experimental results.

Table 6.6. Gait characteristics simulation model and experiment

Characteristic	Simulation	Experimental mean $\pm$ 1 standard deviation
Step length	0.365m	0.37m $\pm$ 0.04m
Step time	0.78s	0.78s $\pm$ 0.09s

Note that the average forward velocity is 0.47 m/s, which is half as fast as human locomotion but faster than many existing humanoid robots.

6.7.2 Motions

The motion of Denise is not a predefined trajectory in time, but rather an autonomous quasi-periodic motion resulting from the natural dynamics of the system. This makes it impossible to compare and average series of steps. Due to small disturbances, each step has a slightly different duration, and so two series will not be synchronized. It would make no sense to present the average motion of multiple experiments, and likewise it is useless to use comparison measures such as mean square error or variance-accounted-for. For two motions that are slightly out of synchronization, these measures would provide low scores even though the walking motions are almost identical. However, we do not necessarily care about trajectories. The reason is that the tracking of trajectories is not on the list of primary requirements for successful walking. A walking motion is successful if it is stable and efficient, if the robots walks a required distance in a required time, and if the robot's gait looks sufficiently appealing. The actual trajectories are not in this list of requirements, and thus they are only of secondary importance. Obviously, for our trust in the ADAMS simulation model we do need to have a good correlation between the simulation and the experiment.

In order to make a useful comparison, we will only look at the trajectory of a single step. In Figs. 6.22–6.25, we have plotted the data of the more than 150 measured steps on top of each other. The start of a step is defined as the instant of heel contact of the leading leg. In this way, we solve the issue of synchronization. The experimental results are not averaged because the result would not represent the actual motions (kinks in the trajectories would be smoothed). We can see that the results from the simulation fall mostly within the variation of the experiments. More importantly, we note that the general shape corresponds well. This means that the discontinuous effects (impacts, controller mode changes) are modeled correctly.

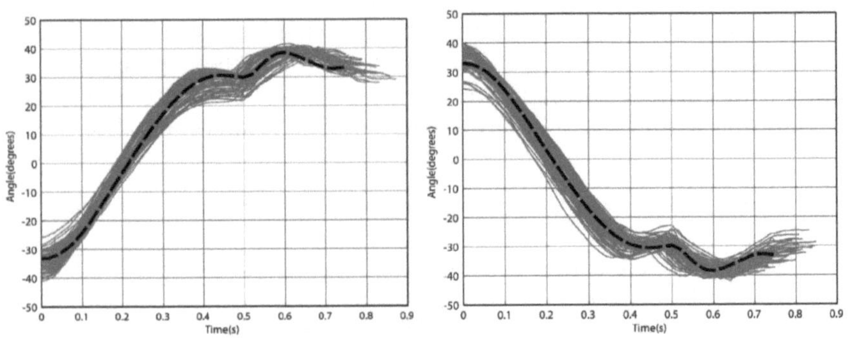

Fig. 6.22. Hip angle between legs. Experiments are grey and the simulation result is plotted using a bold dashed line. **Left:** left leg is stance leg. **Right:** right leg is stance leg.

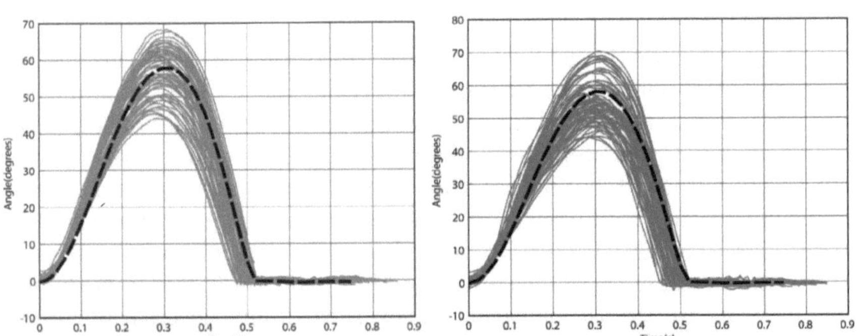

Fig. 6.23. Knee angle. Experiments are grey and the simulation result is plotted using a bold dashed line. **Left:** graph of right knee (left leg is stance leg). **Right:** graph of left knee (right leg is stance leg).

Figures 6.22–6.25 show that the prototype was not fully symmetric; the right knee bends significantly more than the left knee. This can be due to a variety of causes, such as friction in the knee, weaker hip muscles on the left side, or a slight difference in release-behavior of the two knee latches. We believe that

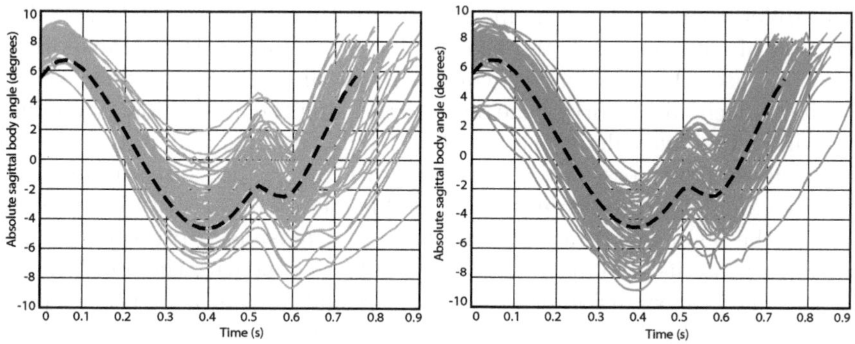

Fig. 6.24. Sagittal motion of the body, measured with respect to the direction of gravity. **Left:** left leg is stance leg. **Right:** right leg is stance leg. The simulation results (bold dashed lines) are equal (symmetric gait), but the experiments (grey lines) indicate a small amount of asymmetry.

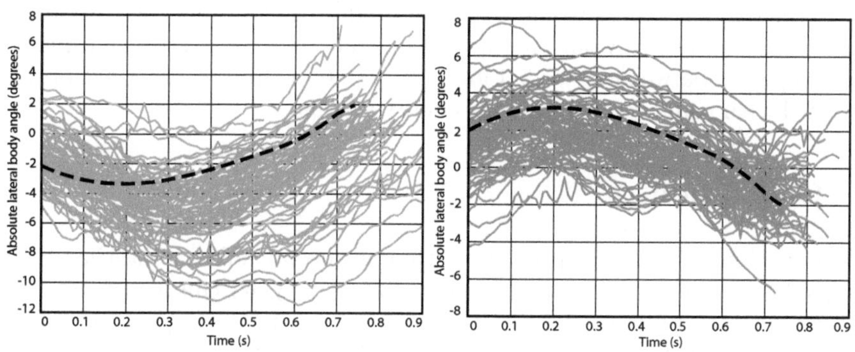

Fig. 6.25. Lateral (sideways) motion of the body, measured with respect to the direction of gravity. **Left:** left leg is stance leg. **Right:** right leg is stance leg. The simulation results (bold dashed lines) are symmetric, but the experiments (grey lines) indicate a small amount of asymmetry. The high-frequency components of the experimental data stem from measurement inaccuracies.

it is currently not necessary to include the asymmetry in the model, given the large random variations introduced by other causes such as those associated with starting the robot walking, which is done by hand.

6.7.3 Energy Consumption

To compare efficiency between humans and robots of different sizes, it is convenient to use the dimensionless specific cost of transport, c_t = (energy used)/(weight x distance traveled). In order to isolate the effectiveness of the mechanical design and controller from the actuator efficiency, we distinguish between the specific energetic cost of transport, c_{et}, and the specific mechanical cost of transport, c_{mt}. Whereas c_{et} uses the total energy consumed by the system, c_{mt} only considers the positive mechanical work of the actuators. Using the muscle forces and length changes in Denise, we calculate $c_{mt} = 0.08$. This does not take into account the huge (but inessential) losses from stepping down the gas pressure. To find a value for c_{et}, we calculated the decrease of available energy (or exergy) for a pressure drop from the 58 atm saturated liquid state to atmospheric pressure. Available energy represents the amount of work that could be done with the pressurized gas if the both the gas expansion process and the simultaneous heat transfer process are reversible (i.e. lossless). In that hypothetical setting, one can use the enthalpy and entropy values for the gas at the beginning and the end of the expansion process. At a constant temperature of 290 K, this amounts to a loss of available energy of 664 kJ per kg CO_2. A 0.45 kg canister can power the 8 kg robot for 30 min of walking at 0.4 m/s yielding $c_{et} = 5.3$. This value has little meaning, however. First, even the best realworld gas-expansion systems can only use about 30% of the theoretically available energy, due to irreversibility issues. More importantly, most of the expansion loss would be eliminated if the CO_2 had been stored at 6 atm. Unfortunately this would require an impractically large storage tank. Thus the discrepancy between $c_{et} = 5.3$ and $c_{mt} = 0.08$ is due to practical problems associated with using compressed-gas energy storage.

6.8 Conclusion

The simulations in this chapter show that a special ankle joint that couples falling sideways (lean) to turning in that direction (yaw) can lead to stable 3D walking models. A practical robustness against disturbances requires a basic form of swing leg control which moves the swing leg quickly to a forward position. With this control rule in place, the model shows behavior that corresponds to bicycles and skateboards; stable motions exist above a certain critical forward velocity, depending on the tilt angle α of the ankle axis. The more vertical the axis (smaller α), the lower the critical velocity. There is a minimum, however; the tilt angle α must always remain larger than the maximal stance leg angle, otherwise it would have a completely vertical orientation at the end of a step. This does not only lead to instabilities but also requires impractically wide feet to prevent tipping over on the inside of the foot.

The ankle joint provides an effective means for direction control; a slight asymmetry in any of the parameters (such as a sideways mass offset) results in a walk on a curved path. The simulations with the elementary model presented in this chapter predict a sufficient robustness against disturbances to warrant the construction of a physical 3D prototype.

Both the elementary simulation model and the prototype demonstrate 3D stability. The skateboard-like ankle joint forms a simple mechanical ingredient for the design of stable dynamic walking bipeds. The stabilizing effect is only present when walking with a substantial forward velocity and requires the presence of hip actuation as proposed in Chapter 4.

Although the idea is strongly linked to a mechanical implementation in the form of a tilted ankle axis, there are many alternative ways of implementation, either mechanically or via control, in the foot, ankle, leg, or hip. The central idea is that, if there exists a forward velocity, a sideways fall can be averted by steering in that direction.

7. Discussion and Future Research Directions

The purpose of this book is to present our work on pneumatic biped robots, so that you may build upon it and avoid our mistakes. Thus, a quintessential part of the information that this book should transmit are our afterthoughts; things that we haven't done (yet) but that we have developed a vision on. These things are presented in this chapter. We hope that these insights will prove useful for future research.

7.1 On General Design Guidelines for Stability

This book has presented a number of design solutions for efficient, natural, and simple two-legged robots. The solutions range from the concept of passive dynamic walking in Chapter 2 to the skateboard-like ankle joint in Chapter 6. One of the most intriguing questions now is how all these solutions fit together, what general ideas lie behind them, and how do they provide stability to the walking motion?

Stability is one of the main issues in bipedal walking because of the fundamental problem of underactuation at the stance foot. The cause of this is the limited size of human-like feet combined with the fact that only compressive forces can be exerted between the foot and the floor. As a result, only limited torques can be transmitted between the foot and the floor, so that there exist two degrees of freedom (fore-aft rotation and sideways lean) that can barely be actuated. This is a cause for instability because the biped system as a whole is operating around the unstable equilibrium of these fundamentally underactuated degrees of freedom; it can be regarded as an inverted pendulum pivoting at the foot.

To understand how stability of the walking motion is achieved, the term 'stability' must be specified more precisely. Walking is a periodic motion in all degrees of freedom (except for forward travel). Therefore, in the phase space (excluding forward travel) the walking motion forms a closed orbit, see Fig. 7.1. The walking motion is stable if there is orbital stability, i.e. if a deviation from the orbit decreases over time. It is important to note that there is a discrete event in the orbit, namely the change-over of foot support (heel strike). Stabilizing effects can

M. Wisse and R.Q. van der Linde: Delft Pneumatic Bipeds, STAR 34, pp. 117–126, 2007.
springerlink.com

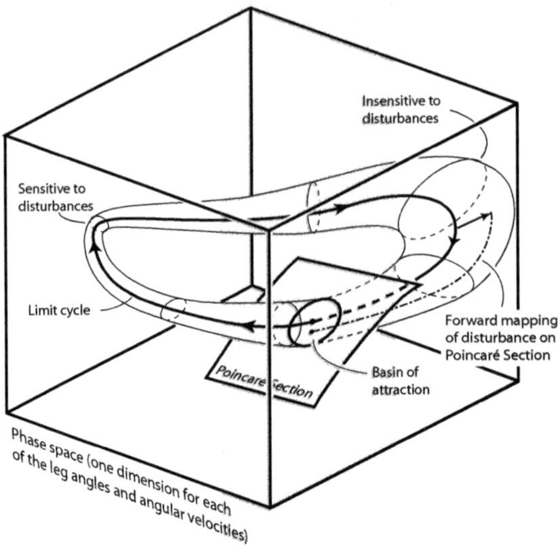

Fig. 7.1. Stylized phase graph of walking motion. The dimensionality of the graph is usually much higher than the three dimensions drawn here, so this figure should be considered only as a sketch to convey the ideas behind the stability analysis of cyclic (walking) motions.

occur during the continuous motion and during the discrete event of heel strike. For example, the 'classical' biped control approach (e.g. Vukobratovic [168]) focuses mainly on stabilizing the orbit during the continuous motion. The robots follow a prescribed trajectory (the continuous part of the orbit) and use control to return to this trajectory after a disturbance. Compare this type of control to a tightrope walker who uses his upper body (plus balancing beam) to stabilize his unstable inverted-pendulum configuration pivoting around the feet. A contrasting example is given with the passive dynamic walkers. Here, orbital stability is mainly obtained from the discrete event at heel strike. The stabilizing effect is the fact that the energy loss at impact is dependent on the forward velocity; if the velocity is too high, more energy is dissipated in the impact and so it provides a stabilizing effect. The stabilizing effect of this impact is stronger than the instability of the inverted-pendulum configuration during the continuous motion. In effect, a small error at the beginning of a step increases during the continuous motion, but the impact strongly brings the motion back to the nominal orbit and thus there is orbital stability. These two extremes (the classical approach versus the concept of passive dynamic walking) illustrate that orbital stability can be realized both by measures during the *continuous* motion and during the *discrete event* at heel strike.

The specific solutions presented in this book can all be seen as contributions to stability in the two categories:

- **Stabilizing effects during continuous motion.** In the search for simple and efficient mechanical solutions for stability, we do not apply the classical solution. The classical solution is to stabilize the unactuated and unstable degrees of freedom at the feet through intensive control of the actuated degrees of freedom (hip and upper body). This solution is control intensive and requires much energy and high-grade components. In contrast, our approach is to allow a mild instability in the degrees of freedom at the foot, and to focus on implementing features that reduce the rate of divergence from the nominal trajectory after a disturbance to an acceptable level:

- Arc feet reduce the rate of divergence; the walker is still an unstable inverted pendulum but deviations from the nominal orbit grow slower than in a walker with point feet.
- The addition of the upper body increases the mass moment of inertia around the foot and thus also reduces the rate of divergence.
- The ankle joint couples yaw to lean and thus the inertia around the vertical axis is added to the lean degree of freedom, which reduces the rate of divergence of this unstable lean degree of freedom. Note that this requires zero slip in the foot-floor contact. It only works as long as a sufficiently large foot-floor contact can be maintained; if the foot would tip over and make ground contact only at a single point, there would be no yaw friction torque between the foot and the floor.
- Most robots in this book are actuated such that they walk at a slightly faster pace than they would if fully passive. As a result, the rate of divergence is slightly decreased *relative to the walking pace.*

In addition to these effects which are all related to the unactuated degrees of freedom at the foot, our design solutions also affect the remaining degrees of freedom. These remaining degrees of freedom, e.g. the hip joint, pose no fundamental instability problem as they can be actuated and controlled directly. The work in this book presents simple and efficient ways to keep these degrees of freedom under control, all in the form of constraints in one form or another:

- The ankle joint couples yaw to lean and so it constrains the yaw degree of freedom; it is no independent degree of freedom and thus it requires no control. Basically, this reduces the dimension of the phase space.
- The bisecting hip mechanism also forms a rigid constraint in such a way that the addition of the upper body does not lead to the addition of any degrees of freedom.
- The knee latches as used in most prototypes lock the knee at the end of the swing phase and all through the stance phase. This is also a reduction of the phase space dimension during an interval at every step.
- The swing leg control as implemented in the prototypes has the effect that the swing leg is quickly brought to a fixed forward position *and then kept there.* By keeping it in that prescribed position at the end of the step, again the phase space dimension is reduced during the final part of the step.

All together, in our robots at the end of the step all degrees of freedom are locked or fixed by control so that only the unactuated degrees of freedom at the foot remain. Hence, all other degrees of freedom do not pose a stability risk.

- **Stabilizing effects during the discrete event at heel strike.** To enable a stabilizing effect during the discrete event at heel strike, it is required that this event actually takes place. In our walker designs, several features have been built in to increase the chances that this occurs:
  - The swing leg control has the effect that the swing leg is *quickly* brought to a forward position. This increases the chances of catching the walker for its next step and prevents a fall forward.
  - The arc feet decrease the chances of falling backward and so they help with ensuring the completion of the step.

  The stabilizing effects during the discrete event at heel strike are the following:
  - The impact at heel strike dissipates more energy when the walking velocity is higher and vice versa, and so it provides a regulating effect.
  - The ankle axis results in a steering motion toward the side that the robot leans to. Effectively, this ensures that the next foothold is in a favorable position so that the discrete event at heel strike puts the walker back on the nominal walking orbit.

One of the directions for future research is to investigate how exactly do features such as the arc feet contribute to the orbital stability. As described above, they reduce the rate of divergence due to the inverted-pendulum instability. However, they might also influence the stabilizing effect of the heel strike impact. These effects were not studied in this book and warrant further research.

To find and investigate new features that can enhance the stability of the walkers, we advise to regard the stability problem as sketched above; stable walking requires orbital stability which can be obtained both during the continuous motion and through the discrete event at heel strike. The walking orbit can be subdivided in various sub-regions, where a disturbance in each of the sub-regions requires a different approach to obtain stability. An example of this can be found in a recent research result on the human reaction to a tripping disturbance [38]. If the disturbance takes place early in the swing phase it is best to use the swing leg muscles to quickly bring it back into its normal forward motion. However, if this disturbance takes place late in the swing phase, it is better to finish the step without undertaking any special corrective action, and just use the stabilizing effects during the discrete event at heel strike. In this manner, it should be possible to obtain highly robust walking machines using only a handful of simple stabilizing features.

7.2 On Stability Measures and Disturbances

Throughout this book, the stability of the walkers is analyzed by means of a Poincaré mapping analysis; how well does the motion respond to changes in the initial conditions of a step? Although this analysis has proven to be both practical and successful, it should be noted that it is only a limited representation of the complete stability problem in dynamic walking robots. The goal of this section is to indicate the limits of the stability analysis as used in this book and to point to future directions of research.

Before discussing the limitations of the stability analysis in this book, it is useful to briefly recapitulate the essence of the cyclic walking motion. Fig. 7.1 shows a stylized phase space graph of a cyclic walking motion. The axes of the graph contain the states of the walker, for example the leg angles and their angular velocities (note that even for the simplest walker, this is already 4-dimensional, which is why Fig. 7.1 is 'stylized' into a 3D figure). A perfect cyclic walking motion (a *limit cycle*) is represented in this graph with a closed orbit which is followed once per stride (two steps). The Poincaré mapping method uses only one point of that orbit; the point where it crosses a predefined surface, the *Poincaré Section*. For walking systems, this surface is usually defined as the instant of heel strike, a condition that occurs only once per step. If the walker is in a limit cycle, the crossing with the Poincaré Section constitutes a point that maps onto itself stride after stride; a *fixed point*. If the walker is not in a limit cycle, it will cross the Poincaré Section at another point. Following the walking motion stride after stride, one will observe a sequence of points in the Poincaré Section which either converges toward a fixed point or diverges until a fall occurs and the sequence stops.

In this book, all stability analyses were performed in the Poincaré Section. We used two forms of analysis, namely a linearized stability analysis of the fixed point, and an analysis of the *basin of attraction*, i.e. the set of all points in the Poincaré Section that converge to the fixed point. The first form, the linearized analysis, has become common for the stability analysis of walking motions since Hurmuzlu [80] first applied it. For small deviations from the fixed point, this analysis provides insight into wether the fixed point (and the corresponding limit cycle) is stable or not, and how fast these small disturbances decrease from stride to stride. In addition to the linearized analysis, this book applies the analysis of the basin of attraction. This analysis provides valuable insight into what happens after larger deviations from the fixed point. We learned to which deviations the walkers were relatively sensitive, and what type of fall would follow. Most importantly, the analysis of the basin of attraction pointed us toward the idea for stabilization by means of hip actuation in Chapter 4.

However, one of the important questions is: is the basin of attraction, or any abstraction thereof, useful as an objective measure for robustness? In this book, we have pretended that it is, first by measuring the total area of the basin of attraction in Chapter 4, and subsequently by measuring (over the principal axes) the smallest distance from the fixed point to the boundaries of the basin of attraction in Chapters 5 and 6. The main objection to the first method, measuring the total area of the basin of attraction, is the fact that it wrongfully attaches value to parts of the basin of attraction far away from the fixed point. For example, assume that the original basin of attraction of a walking motion has the shape of a triangle, and that the fixed point is located somewhere close to the top corner. Then, imagine a design change that results in a downward enlargement of basin of attraction, and simultaneously in an upward shift of the fixed point, bringing it even closer to the top corner. Obviously, the outcome of the area measurement of the basin of attraction wrongfully predicts a better robustness against disturbances.

To avoid the wrongful predictions of the measurement of the area of the basin of attraction, our second approach was to measure the distance from the fixed point to the boundaries of the basin of attraction. Starting in the fixed point, the state variables were varied one by one to find the largest allowable deviations both in positive and in negative direction. In this way, we found the distance between the fixed point and the boundary of the basin of attraction in the direction of the principle axes. After having done this for all the states, the smallest value was taken as a measure for the robustness of the walking motion. Although this measure represented an improvement compared to the area measurement, it is still not an objective measure for robustness. The main objection is that the result of the measure depends on how the state variables have been chosen. One could try to make the comparison between two walking models somewhat fair by applying a standardized scaling (as we have done in all point mass models such as the Simplest Walking Model [54] in Chapter 2) where the models are scaled to unit gravity, leg length and body mass. However, even with such scaling, the outcome of the comparison would still be different if the state variables were chosen differently. In other words, the 'fixed point to boundary'-measurement is not an objective measure for robustness.

So, if the measures for robustness as applied in this book were not objective measures, does there *exist* any objective measure for robustness of the walking motion? To our knowledge, the answer appears to be negative. Following suggestions by Ruina [personal communication], the best approach appears to be to first devise a classification of all possible disturbances that can occur during the walking motion, and then to produce a robustness score for each of the different kinds of disturbances. So, a discussion on the robustness of a certain walking motion should always have a reference to what type of disturbance is considered. Such a classification has not been made in this book, except for the establishment of the 'step-down' experiments in Chapters 4 and 6. A step-down can be regarded as one kind of disturbances for which an objective measure can be created (such as step-down height scaled by leg length). A further classification is recommended for future research, which should incorporate all possible deviations from the limit cycle, and not only those in the Poincaré Section, see Fig. 7.1.

7.3 On Foot Contact in Simulations

The theoretical core of the research is formed by numerical simulations of both irreducibly simple models and more realistic models correlating to the prototypes. Although the basics of multibody modeling and simulation are considered common knowledge, there are some details that have required special attention for the walking models. The most complex issue is the foot contact, especially during heel strike where an impact occurs. Most commercial software packages for multibody simulations cannot be trusted; in one of our attempts the model would take off and fly away at high speed after its first heel strike impact due to a numerical instability. A mild adjustment of the mysterious contact parameters would suddenly result in a more normal motion, but who is to say that that solution is anywhere near reality?

For the development of our simulation equations, it was decided to model the foot contact as completely rigid, which implies the requirement for impact calculations at heel strike. The alternative, 'soft' contact with the use of stiff unilateral springs and dampers underneath the foot, is less attractive due to the additional stiff degrees of freedom to the model. Not only does this cost calculation time during the smooth stance phase, but it is especially disruptive for the stability calculations where all the model states at the beginning of two successive steps have to be compared with each other. Our choice for hard contact with impacts however brings its own peculiarities.

Throughout this book, the models are assumed to have an instantaneous double stance phase; as soon as the swing foot hits the floor at heel contact, the former stance foot loses contact with the floor and does not participate in the actual impact (an assumption which is verified if the rear foot ends up with a positive vertical velocity). Thanks to this instantaneous double stance phase and to the persistent use of arc feet, there is always only one single contact point between the robot and the floor. This makes the impact calculations deterministic and generally solvable. This would not be the case if we would have allowed impacts with multiple simultaneous contacts in our models. The problem is that the impact cannot be regarded as a simple instantaneous event, because the outcome is highly affected by the interaction between the contact forces in the multiple contact points (which are affected by the exact contact stiffness and damping and by how shockwaves travel through the system). We briefly made an excursion toward this more complicated situation in an attempt to capture behaviors such as falling and stopping more realistically. To solve the impact indeterminacies we used an algorithm by Chatterjee [24] which generally provides a likely result in a reasonable amount of time. However, the amount of discussion and uncertainty that result from such calculations are too much of a distraction from the main focus of the research. Therefore, a strong recommendation for future research is to continue to work with the simple impact calculations to avoid the above deliberations, even if it requires mild adjustments to the prototype under investigation.

7.4 On Human Walking

The research in this book was performed on the basis of biomechanical knowledge (e.g. [95, 173, 83]). So, how much do the resulting prototypes resemble a walking human being? The mechanical construction is quite different but the dynamic principles behind it appear to be the same, as revealed by the natural appearance of the motions.

First, the original development of the concept of passive walking by McGeer [108] has been inspired by work of Mochon and McMahon [113] who compared the human swing leg motion with a passive double pendulum. Also the development of the McKibben muscles was inspired by the similarity in compliance with human muscle. Thus, the basic ingredients for the research in this book have a clear relation to human locomotion.

Second, Chapter 4 proposes to obtain fore-aft stability by accelerating the swing leg to move faster than a fully passive leg. The required actuator activity for this corresponds neatly with measurements on muscle activity [83] and with calculations with inverse-dynamic models [95, 142]. Moreover, the required feedback control loop in the prototypes is a simple state machine where a switch underneath the foot signals the transfer to the subsequent state. This can be regarded as a very basic replica of how human walking is governed by a Central Pattern Generator (a neural oscillator in the spinal cord which oscillates in the walking frequency under strong influence of feedback signals). It also correlates with the influential role of foot skin reflexes [162] in human walking.

Third, Chapter 5 proposes the use of a bisecting hip mechanism to obtain a stable upper body. Although the same mechanism can be found in some orthotic devices [81], this is obviously not a good model of the human locomotive system. In our prototypes, the bisecting mechanism is implemented with bicycle chains which could be regarded as tendons or muscles under permanent tension. However, this is a very unattractive solution for biology because of the metabolic energy consumption by the muscle fibers under tension even if no mechanical work is done. As a result, the models and prototypes in this book show relatively large excursions of the upper body whereas the human upper body remains almost vertical. Nevertheless, the prototype results indicate that (at least for 2D motions) a simple and local mechanism (or control loop) around the hip joints already suffices for a stable walking motion; there is no need for more 'system-wide' information such as gyroscopic sensory readings or processed visual information, or even information on which leg is the current stance leg. Therefore, even though the human body does not contain a mechanical bisecting hip joint, it is hypothesized here that the muscles around the hips and pelvic body might very well operate largely on local feedback (reflex) loops alone, as far as the fore-aft motion is concerned.

Finally, Chapter 6 proposes the use of a tilted ankle axis for 3D stability. This axis is mounted in a completely different orientation than in the human ankle, see Fig. 7.2. Nevertheless, the effect is similar; a lean angle to the left results in steering to the left. Again, it is not the physical shape of the design but rather the conceptual idea behind it that could be of interest to the biomechanics research community. The idea is that sideways stability can be obtained by coupling lean to yaw; a potential fall to the right is averted by steering to the right. The success of bicycles and skateboards underlines the power of the concept. It is possible that the human ankle or the entire foot construction contribute to such an effect, but it is equally likely that an active control loop in the hip region induces such an effect. It will be a difficult task to find such effects in

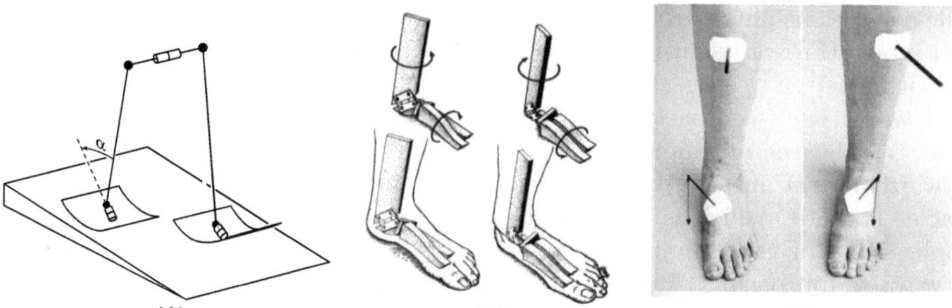

(A) (B) (C)

Fig. 7.2. Robot ankle axis (A) [182] versus human ankle joints (B and C, from Inman [83]). The robot ankle axis points forward and downward whereas the human ankle axis points forward and upward. However, it appears that the overall effect is the same; leaning to the left (or shifting the foot pressure to the left side) results in steering to the left. More research is required for insight into the precise nature of the differences and similarities between the two.

human locomotion, as the human body is always applying multiple solutions simultaneously. In the case of sideways stability, it has been found already that humans apply sideways foot placement as one strong factor of influence and inertial reaction torques (cf. tightrope walking) as the other. This book adds a new possibility; changing the walking direction as a means for sideways stability. It remains as a recommendation for future research in biomechanics to affirm or to negate its use in human locomotion.

7.5 Future Directions

With respect to the current prototypes, it appears that the robustness of the walking motion cannot be increased significantly by means of purely mechanical components. The next logical step will be the inclusion of simple, local reflex-like control loops. We expect that such control loops can help especially with sideways stability; robot research has shown that this is a problem where mechanical solutions alone are not sufficient while there is also biological evidence for the necessity of control to solve this problem. It is known that human beings use sideways foot placement as a control parameter for sideways stability. The sensors with the strongest relation to the sideways stability appear to be the vestibular organ and the pressure distribution sensors underneath the feet. A first research topic could be to find a simple and straightforward control rule that uses these sensors to control the sideways foot placement.

Another issue that might be resolved with local reflex loops is a proper reaction to tripping. Human beings use, depending on the progression of the step, one out of two possible strategies [38], namely 1) quickly setting down the tripping swing foot (shortening the step), or 2) quickly lifting up the tripping swing foot to overcome the obstacle (elongating the step). Possibly, both of these reflexive actions plus the choice between them can be implemented with a simple and straightforward solution (by proper choice of actuator, sensor, and control algorithm).

However, such feed-back stability solutions will probably provide only so much of extra robustness. Human beings obtain much of their robustness from feed-forward control; mostly through visual input an estimate of the upcoming obstacles is made, which is incorporated in advance in the muscle control. The applicability of visual input for walking robot control should be investigated in a two-way approach; 1) How accurate can the upcoming obstacles be predicted with a vision system and what are the key visual cues? 2) What accuracy is required for the predictive information?

As soon as some of the proposed control algorithms are implemented, the issue of tuning the control parameters must be addressed. Manual tuning will be necessary initially but automation is inevitable. In our opinion, the best approach is to provide the system with a set of initial parameter values that are sure to result in stable walking on a smooth floor, probably at the cost of energy expenditure and velocity. Then the system should adapt its control parameters to optimize for stability, efficiency, and velocity. The research challenge is to find the proper adaptation or learning algorithms, and to find which parameters should be tuned in which sequence and on the basis of which measurements.

It will take many man-years of research effort to arrive at a system with sufficient robustness for confidential use in real-world applications. However, we believe that this may happen sooner than one would think; the research community pursuing the approach in this book is quickly gaining momentum with each successful prototype because of the stunningly human impression that these machines leave when strolling along in their natural cadence, while at the same time industrial interest is growing because of the elegant simplicity of the robot designs.

References

1. http://dbl.tudelft.nl. Delft Biorobotics Laboratory.
2. http://ruina.tam.cornell.edu. Research website of Ruina, contains repository of McGeer papers and old walking toy patents.
3. http://www.aerovelco.com. Aerovel Corporation (Tad McGeer).
4. http://www.dynamicwalking.org. Workshop on Dynamic Walking, Michigan, 2006.
5. http://www.humanoid.rise.waseda.ac.jp/. Waseda Humanoid Project.
6. http://www.shadow.org.uk. Shadow Group.
7. http://www.wprize.org. W-prize for 10 km obstacle course, initiated by Tad McGeer.
8. J. Adolfsson, H. Dankowicz, and A. Nordmark. 3D passive walkers: Finding periodic gaits in the presence of discontinuities. *Nonlinear Dynamics*, 24(2):205–229, 2001.
9. P. Artrit and D. G. Caldwell. Single support balancing of a pneumatic biped. In *Proc., Int. Conf. on Climbing and Walking Robots*, pages 835–842, 2001.
10. F. Asano and M. Yamakita. Virtual gravity and coupling control for robotic gait synthesis. *IEEE Trans. on Systems, Man and Cybernetics, Part A*, 31(6):737–745, 2001.
11. C. Assad, S. Trujillo, S. Dastoor, and L. Xu. Cerebellar dynamic state estimation for a biomorphic robot arm. In *Proc., IEEE Int. Conf. on Systems, Man and Cybernetics*, volume 1, pages 877–882, October 2005.
12. C. E. Bauby and A. D. Kuo. Active control of lateral balance in human walking. *Journal of Biomechanics*, 33:1433–1440, 2000.
13. A. Bicchi and G. Tonietti. Fast and soft arm tactics. *IEEE Robotics and Automation Magazine*, 11(2):22–33, 2004.
14. I. Boblan, R. Bannasch, J. Maschuw, A. Schulz, and H. Schwenk. A human-like robot hand and arm with fluidic muscles: Modelling of a muscle driven joint with an antagonistic setup. In *3rd Int. Symposium on Adaptive Motion in Animals and Machines*, Technische Universitt Ilmenau, Germany, September 2005. ISBN 3-938843-03-9.
15. I. Boblan, R. Bannasch, H. Schwenk, L. Miertsch, and A. Schulz. A human like robot hand and arm with fluidic muscles: Biologically inspired construction and functionality. In *Embodied Artificial Intelligence, Dagstuhl Event 03281*, pages 160–179. Springer, 2003.
16. G. A. Borelli. *On the Motion of Animals (De Motu Animalum)*. 1680.

17. E. Borzova and Y. Hurmuzlu. Passively walking five link robot. *Automatica*, 40(4):621–629, 2004.
18. W. Bräune and O. Fischer. *The Human Gait.* Springer-Verlag, Berlin, 1987 (orig. published 1895-1904).
19. D. G. Caldwell, G. A. Medrano-Cerda, and C. J. Bowler. Adaptive position control of antagonistic pneumatic muscle actuators. In *Proc. IEEE Int. Conf. on Intelligent Robots and Systems*, pages 378–384, 1995.
20. D. G. Caldwell, G. A. Medrano-Cerda, and M. Goodwin. Control of pneumatic muscle actuators. *IEEE Control Systems*, pages 40–48, February 1995.
21. D. G. Caldwell, N. Tsagarakis, P. Artrit, and G. A. Medrano-Cerda. Bio-mimetic principles in actuator design for a humanoid robot. *European Journal of Mechanical and Environmental Engineering*, 44(2):75–80, 1999.
22. X. Chang and J. H. Lilly. Fuzzy control for pneumatic muscle tracking via evolutionary tuning. ISSN: 1079-8587.
23. A. Chatterjee. On the realism of complementarity conditions in rigid body collisions. *Nonlinear Dynamics*, 20(2):159–168, 1999.
24. A. Chatterjee and A. Ruina. A new algebraic rigid body collision law based on impulse space considerations. *J. of Applied Mechanics*, 65(4):939–951, 1998.
25. C. Chevallereau, G. Abba, Y. Aoustin, F. Plestan, E. R. Westervelt, C. Canudas-De-Wit, and J. W. Grizzle. RABBIT: a testbed for advanced control theory. *IEEE Control Systems Magazine*, 23(5):57–79, October 2003.
26. C. P. Chou and B. Hannaford. Measurement and modeling of McKibben pneumatic artificial muscles. *IEEE Trans. on Robotics and Automation*, 12(1):90–102, February 1996.
27. C.-P. Chou and B. Hannaford. Study of human forearm posture maintenance with a physiologically based robotic arm and spinal level neural controller. *Biological Cybernetics*, 76:285–298, 1997.
28. R. Colbrunn, G. Nelson, and R. Quinn. Design and control of robotic leg with braided pneumatic actuators. In *Proc., IEEE/RSJ Int. Conf. on Intelligent Robots and Systems*, 2001.
29. R. Colbrunn, G. Nelson, and R. Quinn. Modeling of braided pneumatic actuators for robotic control. In *Proc., IEEE/RSJ Int. Conf. on Intelligent Robots and Systems*, 2001.
30. M. J. Coleman. *A Stability Study of a Three-dimensional Passive-dynamic Model of Human Gait.* PhD thesis, Cornell University, Ithaca, NY, 1998.
31. M. J. Coleman, A. Chatterjee, and A. Ruina. Motions of a rimless spoked wheel: A simple 3D system with impacts. *Dynamics and Stability of Systems*, 12(3):139 – 160, 1997.
32. M. J. Coleman, M. Garcia, K. Mombaur, and A. Ruina. Prediction of stable walking for a toy that cannot stand. *Physical Review E*, 64(2):022901–1 – 022901–3, august 2001.
33. M. J. Coleman and A. Ruina. An uncontrolled toy that can walk but cannot stand still. *Physical Review Letters*, 80(16):3658 – 3661, April 1998.
34. S. H. Collins and A. Ruina. A bipedal walking robot with efficient and human-like gait. In *Proc., IEEE Int. Conf. on Robotics and Automation*, Barcelona, Spain, 2005. IEEE.
35. S. H. Collins, A. Ruina, R. L. Tedrake, and M. Wisse. Efficient bipedal robots based on passive-dynamic walkers. *Science*, 307:1082–1085, February 18 2005.
36. S. H. Collins, M. Wisse, and A. Ruina. A three-dimensional passive-dynamic walking robot with two legs and knees. *Int. J. of Robotics Research*, 20(7):607–615, July 2001.

37. Shadow Robot Company. Developments in dextrous hands for advanced robotic applications. In *Proc., World Automation Congress*, volume 15, pages 123– 128, 2004.
38. A. Forner Cordero. *Human gait, stumble and... fall?* PhD thesis, University of Twente, Enschede, The Netherlands, 2003. ISBN 90-365-1912-8.
39. Bridgestone Corporation. Soft arm acfas robot system, 1987. Tokyo, Japan.
40. Bridgestone Corporation and Taicubo Engineering. Soft boy: Advanced painting system unit, 1993. Tokyo, Japan.
41. F. Daerden and D. Lefeber. Pneumatic artificial muscles: actuators for robotics and automation. *Eur. J. of Mechanical and Environmental Engineering*, 47(1):10–21, 2002.
42. S. Davis and D. G. Caldwell. Braid effects on contractile range and friction modeling in pneumatic muscle actuators. *Int. J. of Robotics Research*, 25(4):359–369, April 2006.
43. S. Davis, N. Tsagarakis, J. Canderle, and D. G. Caldwell. Enhanced modeling and performance in braided pneumatic muscle actuators. *Int. J. of Robotics Research*, 22(3-4):213–227, 2003.
44. S. T. Davis and D. G. Caldwell. The bio-mimetic design of a robot primate using pneumatic artificial muscle actuators. In *Proc., Int. Conf. on Climbing and Walking Robots*, pages 197–204, 2001.
45. J. M. Donelan, R. Kram, and A. D. Kuo. Mechanical and metabolic determinants of the preferred step width in human walking. *Proceedings of the Royal Society of London B*, 268:1985–1992, 2001.
46. J. M. Donelan, D. W. Shipman, R. Kram, and A. D. Kuo. Mechanical and metabolic requirements for active lateral stabilization in human walking. *Journal of Biomechanics*, 37:827–835, 2004.
47. H. Elftman. The measurement of the external force in walking. *Science*, 88:152–153, 1938.
48. G. Endo, J. Morimoto, J. Nakanishi, and G. Cheng. An empirical exploration of a neural oscillator for biped locomotion control. In *Proc., IEEE Int. Conf. on Robotics and Automation*, volume 3, pages 3036 – 3042. IEEE, 2004.
49. T. J Engen and L. F. Ottnat. Upper extremity orthotics: A project report. *Orthopedic and Prosthetic Appliance J.*, pages 112–127, 1967.
50. G. T. Fallis. Walking toy ('improvement in walking toys'). U. S. Patent, No. 376,588, January 17, 1888.
51. D. P. Ferris, K. E. Gordon, G. S. Sawicki, and A. Peethambaran. An improved powered ankle-foot orthosis using proportional myoelectric control. *Gait and Posture*, 23:425–428, 2006.
52. M. Garcia. *Stability, Chaos, and Scaling Laws: Passive-Dynamic Gait Models.* PhD thesis, Cornell University, Ithaca, NY, 1998.
53. M. Garcia, A. Chatterjee, and A. Ruina. Speed, efficiency, and stability of small-slope 2D passive-dynamic bipedal walking. In *Proc., IEEE Int. Conf. on Robotics and Automation*, pages 2351–2356, Piscataway, NJ, 1998.
54. M. Garcia, A. Chatterjee, A. Ruina, and M. J. Coleman. The simplest walking model: Stability, complexity, and scaling. *ASME J. Biomech. Eng.*, 120(2):281–288, April 1998.
55. M. M. Gavrilovic and M. R. Maric. Positional servo-mechanism activated by artificial muscles. *Medical and Biological Eng.*, 7:77–82, 1969.
56. R. H. Gaylord. Fluid actuated motor system and stroking device. Patent 2,844,126, U.S., 1958.

57. T. Geng, B. Porr, and F. Woergoetter. Fast biped walking with a sensor-driven neuronal controller and real-time online learning. *Int. J. of Robotics Research*, 25(3):243–259, 2006.
58. M. W. Gomes and A. Ruina. A walking model with no energy cost. In revision, Phys Rev E, 2005.
59. K. E. Gordon, G. S. Sawicki, and D. P. Ferris. Mechanical performance of artificial pneumatic muscles to power an ankle-foot orthosis. *Journal of Biomechanics*, 39(10):1832–1841, 2006.
60. A. Goswami. Foot-rotation indicator (FRI) point: A new gait planning tool to evaluate postural stability of biped robots. In *Proc., IEEE Int. Conf. on Robotics and Automation*, pages 47–52, Detroit, Michigan, 10-15 May 1999.
61. A. Goswami, B. Espiau, and A. Keramane. Limit cycles and their stability in a passive bipedal gait. In *Proc., IEEE Int. Conf. on Robotics and Automation*, Piscataway, NJ, 1996.
62. A. Goswami, B. Thuilot, and B. Espiau. A study of the passive gait of a compass-like biped robot: symmetry and chaos. *Int. J. Robot. Res.*, 17(12):1282–1301, December 1998.
63. G. Granosik and J. Borenstein. Minimizing air consumption of pneumatic actuators in mobile robots. In *Proc., IEEE Int. Conf. on Robotics and Automation*, pages 3634–3639, 2004.
64. A. A. Grishin, A. M. Formal'sky, A. V. Lensky, and S. V. Zhitomirsky. Dynamic walking of a vehicle with two telescopic legs controlled by two drives. *Int. J. Robotics Research*, 13(2):137–147, 1994.
65. J. W. Grizzle, G. Abba, and F. Plestan. Proving asymptotic stability of a walking cycle for a five dof biped robot model. In *Proc., Int. Conf. Climbing and Walking Robots*, pages 69–81, Portsmouth, UK, Sept 1999.
66. J. W. Grizzle, G. Abba, and F. Plestan. Asymptotically stable walking for biped robots: Analysis via systems with impulse effects. *IEEE Trans. Automat. Contr.*, 46:51–64, Jan 2001.
67. J. J. Grodski and G. B. Immega. Myoelectric control of compliance on a romac protoarm. In *Proc., Int. Symposium on Teleoperation and Control*, pages 297–308, 1988.
68. M. Guihard and P. Gorce. Dynamic control of a biomechanical inspired biped - BIPMAN. In *Proc., Int. Conf. on Climbing and Walking Robots*, pages 39–45, 2001.
69. B. Hannaford and J. M. Winters. *Multiple Muscle Systems*, chapter Actuator properties and movement control: biological and technological models, pages 101–120. Springer-Verlag, New York, 1990.
70. A. H. Hansen, D. S. Childress, and E. H. Knox. Prosthetic foot roll-over shapes with implications for alignment of trans-tibial prostheses. *Prosthetics and Orthotics International*, 24:205–215, 2000.
71. T. Hesselroth, K. Sarkar, P. P. van der Smagt, and K. Schulten. Neural network control of a pneumatic robot arm. *IEEE Transactions on Systems, Man, and Cybernetics*, 24(1):28–38, January 1994.
72. A. Hildebrandt, O. Sawodny, R. Neumann, and A. Hartmann. A flatness based design for tracking control of pneumatic muscle actuators. In *Proc. ICARCV*, volume 3, pages 1156–1161, Singapore, 2002.
73. Hippocrates. *On the Articulations. Parts 52 and 60.* 400 BC. Available at `http://classics.mit.edu//Hippocrates/artic.html`, translated by Francis Adams.

74. K. Hirai, M. Hirose, Y. Haikawa, and T. Takenaka. The development of Honda humanoid robot. In *Proc., IEEE Int. Conf. on Robotics and Automation*, pages 1321–1326, Leuven, Belgium, May 1998.
75. J. K. Hodgins and M. H. Raibert. Adjusting step length for rough terrain locomotion. *IEEE Trans. Robotics and Automation*, 7(3):289 – 298, 1991.
76. J. K. Hodgins and W. L. Wooten. Animating human athletes. In Y. Shirai and S. Hirose, editors, *Robotics Research: The Eighth International Symposium*, pages 356–367, Berlin, 1998. Springer-Verlag.
77. K. Hosoda, T. Takuma, and M. Ishikawa. Design and control of a 3D biped robot actuated by antagonistic pairs of pneumatic muscles. In *Proc., International Symposium on Adaptive Motion in Animals and Machines*, 2005.
78. C. S. Hsu. *Cell-to-cell mapping; a method of global analysis for nonlinear systems.* Applied mathematical sciences 64. New York: Springer, 1987. ISBN 0-387-96520-3.
79. M. Hubbard. Lateral dynamics and stability of the skateboard. *J. of Applied Mechanics*, 46:931–936, December 1979.
80. Y. Hurmuzlu. Dynamics of bipedal gait; part II: Stability analysis of a planar five-link biped. *ASME J. of Applied Mechanics*, 60(2):337–343, 1993.
81. M. J. IJzerman, G. Baardman, H. J. Hermens, P. H. Veltink, H. B. K. Boom, and G. Zilvold. The influence of the reciprocal cable linkage in the advanced reciprocating gait orthosis on paraplegic gait performance. *Prosthetics and orthotics international*, 21:52–61, 1997.
82. Y. Ikemata, A. Sano, and H. Fujimoto. A physical principle of gait generation and its stabilization derived from mechanism of fixed point. In *Proc., IEEE Int. Conf. on Robotics and Automation*, pages 836–841, Orlando, Florida, May 2006.
83. V. T. Inman, H. J. Ralston, and F. Todd. *Human Walking.* Williams & Wilkinns, Baltimore, 1981. ISBN 0-683-04348-X.
84. K. Inoue. Rubbertuators and applications for robots. In *Proc., Int. Symposium on Robotics Research*, pages 57–63, Cambridge, MA, 1988.
85. S. Kajita and K. Tani. Experimental study of biped dynamic walking. *IEEE Control Syst. Mag.*, 16:13–19, Feb 1996.
86. K. Kaneko, F. Kanehiro, S. Kajita, K. Yokoyama, K. Akachi, T. Kawasaki, S. Ota, and T. Isozumi. Design of prototype humanoid robotics platform for HRP. In *Proc., IEEE/RSJ Int. Conf. on Intelligent Robots and Systems, Vol.3*, pages 2431 – 2436, 2002.
87. I. Kato, Y. Mori, and T. Masuda. Pneumatically powered artificial legs walking automatically under various circumstances. In *Proc., Int. Conf. on External Control of Human Extremities*, pages 458–470, 1972.
88. R. Katoh and M. Mori. Control method of biped locomotion giving asymptotic stability of trajectory. *Automatica*, 20(4):405–414, 1984.
89. K. Kawamura, R. A. Peters II, S. Baghi, M. Iskarous, and M. Bishay. Intelligent robotic systems in service of the disabled. *IEEE Trans. on Rehabilitation Engineering*, 1(3):14–21, 1995.
90. T. Kerscher, J. Albiez, and K. Berns. Joint control of the six-legged robot airbug driven by fluidic muscles. In *Proc., Robot Motion and Control*, 2002.
91. T. Kerscher, J. Albiez, J. M. Zoellner, and R. Dillmann. Airinsect - a new innovative biological inspired six-legged walking machine driven by fluidic muscles. In *Proc., 8th Conf. on Intelligent Autonomous Systems*, Amsterdam, The Netherlands, March 2004.
92. T. Kerscher, J. Albiez, J.M. Zoellner, and R. Dillmann. Evaluation of the dynamic model of fluidic muscles using quick-release. In *Proc., IEEE/RAS-EMBS Int. Conf. on Biomedical Robotics and Biomechatronics*, Pisa, Italy, February 2006.

93. D. A. Kingsley and R. D. Quinn. Fatigue life and frequency response of braided pneumatic actuators. In *Proc., IEEE Int. Conf. on Robotics and Automation*, Washington, DC, 2002.
94. G. K. Klute, J. M. Czerniecki, and B. Hannaford. Artificial muscles: Actuators for biorobotic systems. *Int. J. of Robotics Research*, 21(4):295–309, April 2002.
95. H. F. J. M. Koopman. *The three-dimensional analysis and prediction of human walking.* PhD thesis, University of Twente, Enschede, The Netherlands, 1989. ISBN 90-9003075-1.
96. A. D. Kuo. Stabilization of lateral motion in passive dynamic walking. *Int. J. Robot. Res.*, 18(9):917–930, September 1999.
97. A. D. Kuo. Energetics of actively powered locomotion using the simplest walking model. *J. of Biomechanical Engineering*, 124:113–120, February 2002.
98. A. D. Kuo, J. M. Donelan, and A. Ruina. Energetic consequences of walking like an inverted pendulum: step-to-step transitions. *Exercise and sport sciences reviews*, 33(2):88–97, 2005.
99. R. Kurazume, T. Hasegawa, and K. Yoneda. The sway compensation trajectory for a biped robot. In *Proc., IEEE Int. Conf. on Robotics and Automation*, pages 925 – 931, 2003. Vol. 1.
100. Y. Kuroki, M. Fujita, T. Ishida, K. Nagasaka, and J. Yamaguchi. A small biped entertainment robot exploring attractive applications. In *Proc., IEEE Int. Conf. on Robotics and Automation*, pages 471–476, Taipei, Taiwan, 14-19 September 2003.
101. Y. K. Lee and I. Shimoyama. A skeletal framework artificial hand actuated by pneumatic artificial muscles. In *Proc., IEEE Int. Conf. on Robotics and Automation*, volume 2, pages 926–931, 1999.
102. S. Lightner. The fluidic muscle: A 'new' development. *Journal of Modern Engineering*, 2(2), 2002.
103. E.-J. Marey. *Le Mouvement.* 1894.
104. R. Margaria. *Biomechanics and Energetics of Muscular Exercise.* Clarendon Press, Oxford, U.K., 1976.
105. N. M. Mayer, A. A. F.-Nassiraei, F. Farkas, Z. Hsu, and T. Christaller. Stabilizing dynamic walking with physical tricks. In *Proc., Int. Conf. Climbing and Walking Robots*, Madrid, Spain, September 2004.
106. T. McGeer. Powered flight, child's play, silly wheels, and walking machines. In *Proc., IEEE Int. Conf. on Robotics and Automation*, pages 1592–1597, Piscataway, NJ, 1989.
107. T. McGeer. Passive bipedal running. *Proc. of the Royal Society of London: Biological Sciences*, 240:107–134, 1990.
108. T. McGeer. Passive dynamic walking. *Int. J. Robot. Res.*, 9(2):62–82, April 1990.
109. T. McGeer. Passive walking with knees. In *Proc., IEEE Int. Conf. on Robotics and Automation*, pages 1640–1645, Los Alamitos, CA, 1990.
110. T. McGeer. Passive dynamic biped catalogue. In R. Chatila and G. Hirzinger, editors, *Proc., Experimental Robotics II: The 2nd International Symposium*, pages 465–490, Berlin, 1991. Springer-Verlag.
111. T. McGeer. *Mechanics of Animal Locomotion*, volume 11 of *Advances in Comparative and Environmental Physiology*, chapter Principles of Walking and Running. Springer-Verlag, Berlin, 1992.
112. T. McGeer. Dynamics and control of bipedal locomotion. *J. theor. Biol.*, 163:277–314, 1993.
113. S. Mochon and T. A. McMahon. Ballistic walking. *J. Biomechanics*, 13:49–57, 1980.

114. J. Morimoto, G. Cheng, C. G. Atkeson, and G. J. Zeglin. A simple reinforcement learning algorithm for biped walking. In *Proc., IEEE Int. Conf. on Robotics and Automation*, volume 3, pages 3030–3035, 2004.
115. J. Morimoto, J. Nakanishi, G. Endo, and G. Cheng. Acquisition of a biped walking pattern using a poincare map. In *Proc., IEEE/RAS Int. Conf. on Humanoid Robots*, pages 912–924, Nov 2004.
116. J. Morimoto, J. Nakanishi, G. Endo, G. Cheng, C. G. Atkeson, and G. J. Zeglin. Poincar-map-based reinforcement learning for biped walking. In *Proc., IEEE Int. Conf. on Robotics and Automation*, pages 2381–2386, 2005.
117. J. Morimoto, G. J. Zeglin, and C. G. Atkeson. Minimax differential dynamic programming: application to a biped walking robot. In *Proc., Int. Conf. on Intelligent Robots and Systems*, volume 2, pages 1927 – 1932. IEEE, 2003.
118. E. Muybridge. *The Human Figure in Motion.* Dover Publications, New York, 1955, 1985, 1995. ISBN 0486202046.
119. N. Nadjar-Gauthier, H. Cherrid, and J.-C. Cadiou. A new second order sliding mode control for the experimental walking of an electro-pneumatic biped robot. In *Proc., Int. Conf. on Climbing and Walking Robots*, pages 93–100, 2002.
120. J. Nakanishi, J. Morimoto, G. Endo, G. Cheng, S. Schaal, and M. Kawato. Learning from demonstration and adaptation of biped locomotion. *Robotics and Autonomous Systems*, 47(79-91), 2004.
121. V. L. Nickel, M. D. J. Perry, and A. L. Garret. Development of useful function in the severely paralysed hand. *J. Bone Joint Surgery*, 45A(5):933–952, 1963.
122. M. Okada, T. Shinohara, T. Gotoh, S. Ban, and Y. Nakamura. Double spherical joint and backlash clutch for lower limbs of humanoids. In *Proc., IEEE Int. Conf. on Robotics and Automation*, pages 491–496, 2003.
123. K. Ono, R. Takahashi, and T. Shimada. Self-excited walking of a biped mechanism. *Int. J. of Robotics Research*, 20(12):953–966, 2001.
124. K. Ono, F. Takasahi, and R. Takahashi. Self-excited walking of a biped mechanism with feet. *Int. J. of Robotics Research*, 23(1):55–68, 2004.
125. R. T. Pack, J. L. Christopher Jr., and K. Kawamura. A rubbertuator-based structure-climbing inspection robot. In *Proc., IEEE Int. Conf. on Robotics and Automation*, pages 1869–1874, Albuquerque, NM, 1997.
126. N-C. Park, H-W. Park, H-S. Yang, and Y-P. Park. Robust position and force control of two D.O.F. flexible manipulators with artificial muscle. In *Proc. Int. Conf. Motion and Vibration Contr.*, Zurich, Switzerland, 1998.
127. P. T. Piiroinen. *Recurrent Dynamics of Nonsmooth Systems with Application to Human Gait.* PhD thesis, Royal Institute of Technology, Stockholm, Sweden, 2002.
128. D. H. Plettenburg. *Upper Extremity Prosthetics; Current status & evaluation.* VSSD, Delft, 2006. ISBN-13 978-90-71301-75-9. E-book version: ISBN-13 978-90-71301-77-3.
129. J. E. Pratt. *Exploiting inherent robustness and natural dynamics in the control of bipedal walking robots.* PhD thesis, Massachusetts Institute of Technology, Cambridge, MA, 2000.
130. J. E. Pratt, C.-M. Chew, A. Torres, P. Dilworth, and G. Pratt. Virtual model control: An intuitive approach for bipedal locomotion. *Int. J. of Robotics Research*, 20(2):129–143, 2001.
131. IEEE/RAS Int. conf. on Humanoid Robots Proc., editor. *Passive dynamic autonomous control of bipedal walking*, volume 2, Nov 2004.
132. M. H. Raibert. *Legged robots that balance.* The MIT Press, Cambridge, Massachusetts, 1986. ISBN 0-262-18117-7.

133. H. J. Ralston. Energy-speed relation and optimal speed during level walking. *Int. z. angew. Physiol.*, 17:277–283, 1958.
134. R. Regele, W. Bott, and P. Levi. Prorobot - predictions for the future development of humanoid robots. In R. Dillmann, H. Wörn, and T. Gockel, editors, *Autonome Mobile Systeme*, pages 292–303. Springer, Berlin, 2003.
135. C. Sabourin, O. Bruneau, and G. Buche. Control strategy for the robust dynamic walk of a biped robot. *Int. J. Robotics Research*, 25(9):843–860, 2006.
136. Y. Sakagami, R. Watanabe, C. Aoyama, S. Matsunaga, N. Higaki, and M. Fujita. The intelligent ASIMO: System overview and integration. In *Proc., Int. Conf. on Intelligent Robots and Systems*, pages 2478–2483, Lausanne, Switzerland, 30 September - 4 October 2002.
137. P. Scarfe and E. Lindsay. Air muscle actuated low cost humanoid hand. *International Journal of Advanced Robotic Systems*, 3(1):139–146, 2006.
138. H. F. Schulte. *The application of external power in prosthetics and orthotics*, chapter The characteristics of the McKibben artificial muscle, pages 94–115. Number publication 874. National Academy of Sciences, Washington DC, 1961.
139. A. L. Schwab and J. P. Meijaard. Dynamics of flexible multibody systems with non-holonomic constraints: A finite element approach. *Multibody System Dynamics*, 10(12):107–123, 2003.
140. A. L. Schwab, J. P. Meijaard, and J. M. Papadopoulos. Benchmark results on the linearized equations of motion of an uncontrolled bicycle. *KSME Journal of Mechanical Science and Technology*, 19(1):292–304, 2005.
141. A. L. Schwab and M. Wisse. Basin of attraction of the simplest walking model. In *Proc., ASME Design Engineering Technical Conferences*, Pittsburgh, Pennsylvania, 2001. ASME. Paper number DETC2001/VIB-21363.
142. R. Selles, J. B. J. Bussmann, R. C. Wagenaar, and H. J. Stam. Comparing predictive validity of four ballistic swing phase models of human walking. *J. of Biomechanics*, 34:1171–1177, 2001.
143. M. W. Spong and F. Bullo. Controlled symmetries and passive walking. In *IFAC Triennial World Congress*, Barcelona, Spain, 2002.
144. T. Takuma, K. Hosoda, and M. Asada. Walking stabilization of biped with pneumatic actuators against terrain changes. In *Proc., IEEE/RSJ International Conference on Intelligent Robots and Systems*, pages 2775–2780, 2005.
145. T. Takuma, K. Hosoda, M. Ogino, and M. Asada. Controlling walking period of a pneumatic muscle walker. In *International Conference on Climbing and Walking Robots*, 2004.
146. T. Takuma, K. Hosoda, M. Ogino, and M. Asada. Stabilization of quasi-passive pneumatic muscle walker. In *IEEE-RAS/RSJ International Conference on Humanoid Robots*, volume 2, pages 627– 639, Los Angeles, USA, November 2004.
147. T. Takuma, S. Nakajima, K. Hosoda, and M. Asada. Design of self-contained biped walker with pneumatic actuators. In *SICE Annual Conference*, 2004. WPI-2-4.
148. K. Tanie. Humanoid robot and its application possibility. In *Proc., IEEE Int. Conf. on Multisensor Fusion and Integration for Intelligent Systems*, pages 213–214, 2003.
149. R. Tedrake, T. W. Zhang, and H. S. Seung. Stochastic policy gradient reinforcement learning on a simple 3D biped. In *Proc., IEEE Int. Conf. on Intelligent Robots and Systems*, pages 2849–2854, 2004.
150. R. Tedrake, T.W. Zhang, M.-F. Fong, and H.S. Seung. Actuating a simple 3D passive dynamic walker. In *Proc., IEEE Int. Conf. on Robotics and Automation*, 2004.

151. B. Tondu, S. Ippolito, and J. Guiochet. A seven-degrees-of-freedom robot-arm driven by pneumatic artificial muscles for humanoid robots. *Int. J. of Robotics Research*, 24(4):257–274, April 2005.
152. B. Tondu and P. Lopez. Modeling and control of mckibben artificial muscle robot actuators. *IEEE Control Systems Magazine*, pages 15–38, April 2000.
153. G. Tonietti and A. Bicchi. Adaptive simultaneous position and stiffness control for a soft robot arm. In *Proc. IEEE Int. Conf. on Intelligent Robots and Systems*, pages 1992–1997, 2002.
154. N. G. Tsagarakis and D. G. Caldwell. Development and control of a "soft-actuated" exoskeleton for use in physiotherapy and training. *Autonomous Robots (special issue on rehabilitation robotics)*, 15(1):21–33, 2003.
155. P. Tuffield and H. Elias. The shadow robot mimics human actions. *Industrial Robot*, 30(1):56–60, 2003.
156. G. J. M. Tuijhof and J. L. Herder. Design, actuation and control of an anthropomorphic robot arm. *Mechanism and Machine Theory*, 35:945–962, 2000.
157. R. Q. van der Linde. Active leg compliance for passive walking. In *Proc., IEEE Int. Conf. on Robotics and Automation*, pages 2339–2344, Leuven, Belgium, 16-20 May 1998.
158. R. Q. van der Linde. Design, analysis and control of a low power joint for walking robots, by phasic activation of mckibben muscles. *IEEE Trans. Robotics and Automation*, 15(4):599–604, August 1999.
159. R. Q. van der Linde. Passive bipedal walking with phasic muscle contraction. *Biological Cybernetics*, 81(3):227–237, September 1999.
160. R. Q. van der Linde. *Bipedal walking with active springs, gait synthesis and prototype design.* PhD thesis, Delft University of Technology, Delft, The Netherlands, November 2001. ISBN 90-370-0193-9.
161. P. P. van der Smagt, F. C. A. Groen, and K. Schulten. Analysis and control of a rubbertuator arm. *Biological Cybernetics*, 75(5):433–440, 1996.
162. B. M. H. van Wezel, F. A. M. Ottenhoff, and J. Duysens. Dynamic control of location-specific information in tactile cutaneous reflexes from the foot during human walking. *J. of Neuroscience*, 17(10):3804–3814, 1997.
163. B. Vanderborght, B. Verrelst, R. Van Ham, and D. Lefeber. Controlling a bipedal walking robot actuated by pleated pneumatic artificial muscles. *Robotica*, 24(4):401–410, 2006.
164. J. Vermeulen, B. Verrelst, B. Vanderborght, and D. Lefeber. Trajectory planning for the walking biped 'Lucy'. *Int. J. of Robotics Research*, 25(9):867–887, 2006.
165. B. Verrelst, R. Van Ham, B. Vanderborght, F. Daerden, and D. Lefeber. The pneumatic biped 'Lucy' actuated with pleated pneumatic artificial muscles. *Autonomous Robots*, 18:201–213, 2005.
166. B. Verrelst, R. Van Ham, B. Vanderborght, D. Lefeber, F. Daerden, and M. Van Damme. Second generation pleated pneumatic artificial muscle and its robotic applications. *Advanced Robotics*, 20(7):783–805, 2006.
167. B. Verrelst, J. Vermeulen, B. Vanderborght, R. Van Ham, J. Naudet, D. Lefeber, F. Daerden, and M. Van Damme. Motion generation and control for the pneumatic biped 'Lucy'. *Int. J. of Humanoid Robotics*, 3(1):1–35, 2006.
168. M. Vukobratovic. How to control the artificial anthropomorphic systems. *IEEE Trans. System, Man and Cybernetics SMC-3*, pages 497–507, 1973.
169. M. Vukobratovic and B. Borovac. Zero-moment point - thirty five years of its life. *Int. J. of Humanoid Robotics*, 1(1):157–173, March 2004.
170. W. Weber and E. Weber. *Mechanik der Menschlichen Gehwerkzeuge: eine anatomisch-physiologische Untersuchung.* Gottingen, 1836.

171. E. R. Westervelt, J. W. Grizzle, and D. E. Koditschek. Hybrid zero dynamics of planar biped walkers. *IEEE Trans. on Automatic Control*, 48(1):42–56, 2003.
172. J. E. Wilson. Walking toy. U. S. Patent, No. 2,140,275, October 15 1938.
173. D. A. Winter. *Biomechanics and motor control of human movement.* John Wiley & Sons, Inc., 1990. ISBN 0-471-50908-6.
174. J. M. Winters and E. S. Sagranichiny. Why braided pneumatic actuators in rehabilitation robotics? principles, properties, and suggested applications. In *Proc., Int. Conf. Rehabilitation Robotics*, pages 201–208, Wilmington, 1994.
175. M. Wisse. Passive dynamic biped with knees. Master's thesis, Delft University of Technology, Delft, The Netherlands, June 2000.
176. M. Wisse. Three additions to passive dynamic walking: actuation, an upper body, and 3D stability. *Int. J. of Humanoid Robotics*, 2(4):459–478, December 2005.
177. M. Wisse, C. G. Atkeson, and D. K. Kloimwieder. *Dynamic stability of a simple biped walking system with swing leg retraction.* Springer, October 2006. ISBN: 3540361189.
178. M. Wisse, G. Feliksdal, J. van Frankenhuyzen, and B. Moyer. Passive-based robot denise, a simple, efficient, and lightweight biped. *IEEE Robotics and Automation Magazine*, 2007. In press.
179. M. Wisse, D. G. E. Hobbelen, R. J. J. Rotteveel, S. O. Anderson, and G. J. Zeglin. Ankle springs instead of arc-shaped feet for passive dynamic walkers. In *Proc., IEEE/RAS Int. Conf. on Humanoid Robots*, Genoa, Italy, December 2006.
180. M. Wisse, D. G. E. Hobbelen, and A. L. Schwab. Adding the upper body to passive dynamic walking robots by means of a bisecting hip mechanism. *IEEE Trans. on Robotics*, 23(1):112–123, 2007.
181. M. Wisse and A. L. Schwab. A 3D passive dynamic biped with roll and yaw compensation. *Robotica*, 19:275–284, 2001.
182. M. Wisse and A. L. Schwab. Skateboards, bicycles, and 3D biped walkers; velocity dependent stability by means of lean-to-yaw coupling. *Int. J. of Robotics Research*, 24(6):417–429, 2005.
183. M. Wisse, A. L. Schwab, and F. C. T. van der Helm. Passive dynamic walking model with upper body. *Robotica*, 22:681–688, 2004.
184. M. Wisse, A. L. Schwab, R. Q. van der Linde, and F. C. T. van der Helm. How to keep from falling forward; elementary swing leg action for passive dynamic walkers. *IEEE Trans. on Robotics*, 21(3):393–401, 2005.
185. M. Wisse and J. van Frankenhuyzen. Design and construction of Mike; a 2D autonomous biped based on passive dynamic walking. In *Proc., Conference on Adaptive Motion of Animals and Machines, AMAM*, Kyoto, Japan, 2003. Paper number WeP-I-1.
186. M. Wisse and J. van Frankenhuyzen. Design and construction of Mike; a 2D autonomous biped based on passive dynamic walking. In H. Kimura and K. Tsuchiya, editors, *Adaptive Motion of Animals and Machines*, pages 143–154, Tokyo, 2006. Springer-Verlag.

Springer Tracts in Advanced Robotics

Edited by B. Siciliano, O. Khatib and F. Groen

Vol. 34: Wisse, M.; van der Linde, R.Q.
Delft Pneumatic Bipeds
136 p. 2007 [978-3-540-72807-8]

Vol. 33: Kong, X.; Gosselin, C.
Type Synthesis of Parallel
Mechanisms
272 p. 2007 [978-3-540-71989-2]

Vol. 32: Milutinović, D.; Lima, P.
Cells and Robots: Modeling and Control of
Large-Size Agent Populations
130 p. 2007 [978-3-540-71981-6]

Vol. 31: Ferre, M.; Buss, M.; Aracil, R.;
Melchiorri, C.; Balaguer C. (Eds.)
Advances in Telerobotics
500 p. 2007 [978-3-540-71363-0]

Vol. 30: Brugali, D. (Ed.)
Software Engineering for Experimental Robotics
490 p. 2007 [978-3-540-68949-2]

Vol. 29: Secchi, C.; Stramigioli, S.; Fantuzzi, C.
Control of Interactive Robotic Interfaces – A
Port-Hamiltonian Approach
225 p. 2007 [978-3-540-49712-7]

Vol. 28: Thrun, S.; Brooks, R.; Durrant-Whyte, H.
(Eds.)
Robotics Research – Results of the 12th
International Symposium ISRR
602 p. 2007 [978-3-540-48110-2]

Vol. 27: Montemerlo, M.; Thrun, S.
FastSLAM – A Scalable Method for the
Simultaneous Localization and Mapping
Problem in Robotics
120 p. 2007 [978-3-540-46399-3]

Vol. 26: Taylor, G.; Kleeman, L.
Visual Perception and Robotic Manipulation – 3D
Object Recognition, Tracking and Hand-Eye
Coordination
218 p. 2007 [978-3-540-33454-5]

Vol. 25: Corke, P.; Sukkarieh, S. (Eds.)
Field and Service Robotics – Results of the 5th
International Conference
580 p. 2006 [978-3-540-33452-1]

Vol. 24: Yuta, S.; Asama, H.; Thrun, S.;
Prassler, E.; Tsubouchi, T. (Eds.)
Field and Service Robotics – Recent Advances in
Research and Applications
550 p. 2006 [978-3-540-32801-8]

Vol. 23: Andrade-Cetto, J,; Sanfeliu, A.
Environment Learning for Indoor Mobile Robots
– A Stochastic State Estimation Approach
to Simultaneous Localization and Map Building
130 p. 2006 [978-3-540-32795-0]

Vol. 22: Christensen, H.I. (Ed.)
European Robotics Symposium 2006
209 p. 2006 [978-3-540-32688-5]

Vol. 21: Ang Jr., H.; Khatib, O. (Eds.)
Experimental Robotics IX – The 9th International
Symposium on Experimental Robotics
618 p. 2006 [978-3-540-28816-9]

Vol. 20: Xu, Y.; Ou, Y.
Control of Single Wheel Robots
188 p. 2005 [978-3-540-28184-9]

Vol. 19: Lefebvre, T.; Bruyninckx, H.;
De Schutter, J. Nonlinear Kalman Filtering
for Force-Controlled Robot Tasks
280 p. 2005 [978-3-540-28023-1]

Vol. 18: Barbagli, F.; Prattichizzo, D.;
Salisbury, K. (Eds.)
Multi-point Interaction with Real
and Virtual Objects
281 p. 2005 [978-3-540-26036-3]

Vol. 17: Erdmann, M.; Hsu, D.; Overmars, M.;
van der Stappen, F.A (Eds.)
Algorithmic Foundations of Robotics VI
472 p. 2005 [978-3-540-25728-8]

Vol. 16: Cuesta, F.; Ollero, A.
Intelligent Mobile Robot Navigation
224 p. 2005 [978-3-540-23956-7]

Vol. 15: Dario, P.; Chatila R. (Eds.)
Robotics Research – The Eleventh
International Symposium
595 p. 2005 [978-3-540-23214-8]

Vol. 14: Prassler, E.; Lawitzky, G.; Stopp, A.; Grunwald, G.; Hägele, M.; Dillmann, R.; Iossifidis. I. (Eds.)
Advances in Human-Robot Interaction
414 p. 2005 [978-3-540-23211-7]

Vol. 13: Chung, W.
Nonholonomic Manipulators
115 p. 2004 [978-3-540-22108-1]

Vol. 12: Iagnemma K.; Dubowsky, S.
Mobile Robots in Rough Terrain – Estimation, Motion Planning, and Control with Application to Planetary Rovers
123 p. 2004 [978-3-540-21968-2]

Vol. 11: Kim, J.-H.; Kim, D.-H.; Kim, Y.-J.; Seow, K.-T.
Soccer Robotics
353 p. 2004 [978-3-540-21859-3]

Vol. 10: Siciliano, B.; De Luca, A.; Melchiorri, C.; Casalino, G. (Eds.)
Advances in Control of Articulated and Mobile Robots
259 p. 2004 [978-3-540-20783-2]

Vol. 9: Yamane, K.
Simulating and Generating Motions of Human Figures
176 p. 2004 [978-3-540-20317-9]

Vol. 8: Baeten, J.; De Schutter, J.
Integrated Visual Servoing and Force Control – The Task Frame Approach
198 p. 2004 [978-3-540-40475-0]

Vol. 7: Boissonnat, J.-D.; Burdick, J.; Goldberg, K.; Hutchinson, S. (Eds.)
Algorithmic Foundations of Robotics V
577 p. 2004 [978-3-540-40476-7]

Vol. 6: Jarvis, R.A.; Zelinsky, A. (Eds.)
Robotics Research – The Tenth International Symposium
580 p. 2003 [978-3-540-00550-6]

Vol. 5: Siciliano, B.; Dario, P. (Eds.)
Experimental Robotics VIII – Proceedings of the 8th International Symposium ISER02
685 p. 2003 [978-3-540-00305-2]

Vol. 4: Bicchi, A.; Christensen, H.I.; Prattichizzo, D. (Eds.)
Control Problems in Robotics
296 p. 2003 [978-3-540-00251-2]

Vol. 3: Natale, C.
Interaction Control of Robot Manipulators – Six-degrees-of-freedom Tasks
120 p. 2003 [978-3-540-00159-1]

Vol. 2: Antonelli, G.
Underwater Robots – Motion and Force Control of Vehicle-Manipulator Systems
268 p. 2006 [978-3-540-31752-4]

Vol. 1: Caccavale, F.; Villani, L. (Eds.)
Fault Diagnosis and Fault Tolerance for Mechatronic Systems – Recent Advances
191 p. 2003 [978-3-540-44159-5]

MIX
Papier aus verantwortungsvollen Quellen
Paper from responsible sources
FSC® C105338

If you have any concerns about our products,
you can contact us on
ProductSafety@springernature.com

In case Publisher is established outside the EU,
the EU authorized representative is:
Springer Nature Customer Service Center GmbH
Europaplatz 3, 69115 Heidelberg, Germany

Printed by Libri Plureos GmbH
in Hamburg, Germany